JN409078

ARCTIC AND ANTARCTIC SEA ICE

Kyungsik Choi

Dasom Publishing Co.

Preface

The effects of global warming evidenced in the thinning and decrease in the mass of the polar sea ice are now old news. In order to provide foundational information for the design of new icebreaking vessels and marine structures that could navigate such changing seascape, the need for research on sea ice becomes more pressing. The ARAON, Korea's first icebreaking research vessel, has been used for scientific research activities in the Arctic and Antarctic seas every year since its construction in 2009. The field test on board the ARAON encompasses the icebreaking test to verify the correlation between the model test performed in the ice model basin and the ice field test with an actual ship. In the ice field test, material properties of sea ice are measured and the glacial load caused by collisions between ship and ice is estimated. Since it is possible to estimate the ice load more accurately through the correlation analysis between the icebreaking performance of the ice field test and the model test results, it can be said that the ice field test is an essential procedure for the design of an icebreaking vessel.

Recently, various polar studies using the KRISO ice model basin and the icebreaking research vessel ARAON are being conducted, and scientific and technological interest in polar region is increasing at home and abroad. With these in mind, the first edition was published under the title of 'Sea Ice in Polar Regions' (GeoBook, 2012), and was prepared in the purpose of broadening the basis of domestic ice mechanics and polar engineering researches. After the first edition, the need for an English edition was raised. For the new English edition, the contents with new photos and updated information were supplemented.

This book is a compilation of photos of various ice features taken by the author during the four trips between 2010 and 2019 he took to carry out ice field tests in the Arctic and Antarctic seas on board the ARAON. In addition, photos of sea ice in the Baltic Sea, Finland, sea ice in Svalbard, Norway, and glaciers in Alaska, USA, taken during an ice mechanics experiment in 2010, are also included.

It is hoped that this book will serve as a reference book for practitioners who are involved in polar engineering as well as an introductory book for the general public who are interested in icebreaking ships and sea ice. Because Korea is not geographically close to polar regions, securing vivid data seen and felt in the cold environments is the biggest difficulty in expanding the basis of polar engineering research. It is expected that this book will be a very valuable resource for us. Many people have helped to bring this book to the light. Above all, I would like to thank the Korea Polar Research Institute for their cooperation in the use of the ARAON for my research. I would also like to express my gratitude to the many researchers who were on board with the author and performed the ice field test together, whose expertise and fellowship had been the cornerstones in all parts of its process.

September 2021

Kyungsik Choi

Contents

II

Introduction to Polar Regions and Sea Ice

Ice Mechanics and Polar Engineering

The Arctic region refers to the north of the Arctic Circle (66°N 33') and geographically consists of the Arctic Ocean surrounded by Siberia, Northern Europe, Canada and Alaska. Climatologists define the Arctic as region where the average temperature of the warmest month of the year is below 10°C and the average annual temperature is below 0°C. This boundary coincides with the northern limit of tree line and roughly coincides with the southern limit of permafrost. Oceanographers also consider the southernmost limit line of ice drift to be the Arctic Circle. The continental shelf of the Arctic Ocean is rich in oil and natural gas, and oil production has been taking place in these regions since the late 1970s. However, since the Arctic Ocean is always covered with thick ice except for a short period of summer, in order to develop resources in the polar region, special equipment such as an icebreaker capable of overcoming cold temperatures and sea ice and ice mechanics/polar engineering technology are required.

▲
ARAON's icebreaking performance test in the Antarctic sea ice (Courtesy of KOPRI)

▲ Measuring sea ice properties on the Arctic sea ice

Antarctica is a large continent with an area of 13.6 million square km, and most of the land is covered with ice cap with an average thickness of 2,160 m. Antarctica is geographically opposite to that of the Arctic, but as a continent surrounded by sea, the wind is very strong. In addition, Antarctica is a place with very little annual precipitation, so the natural circulation of substances is very slow with low temperature. Antarctica also has abundant biological and mineral resources in and around the continent, but in order to preserve the geographical importance and natural environment of Antarctica, the Antarctic Treaty, which took effect in 1961, freezes all territorial rights and prohibits development. However, the unique natural environment that Antarctica preserves is playing an important role as a natural science laboratory that studies the creation and change of the earth.

Navigators have suffered a lot from ice when navigating in icy seas, and civil engineers have been preoccupied with constructing bridges across ice-covered rivers or structures in frozen land. Ice mechanics is the product of an effort to scientifically understand the relationship between man-made structures and ice. Activities to develop oil and natural gas reserves, along the coast of the Arctic Ocean, have resulted in a lot of new knowledge about ice mechanics. Ice is a very unique material that usually cracks when subjected to force, but when force is applied very slowly or at temperatures close to freezing point, it becomes a continuum rather than a brittle material.

The most important problem in ice mechanics is how much of the ice load can be tolerated by the structures to be installed in the ice infested area when they come in contact with ice. The strength of ice differs several tens of times between that measured in the laboratory with a small ice specimen and in the actual ice field, so it is very difficult to apply the data obtained in the laboratory to the actual ice field. This is the reason why the ice field test is essential. Another interesting fact is that when constructing a structure, we are interested in how strong the structure is, that is, how much load the material can withstand. But in ice mechanics, we approach the problem from a different perspective, that is, how easily the ice can be destroyed.

▾ The KRISO ice model basin in Korea (Courtesy of KRISO)

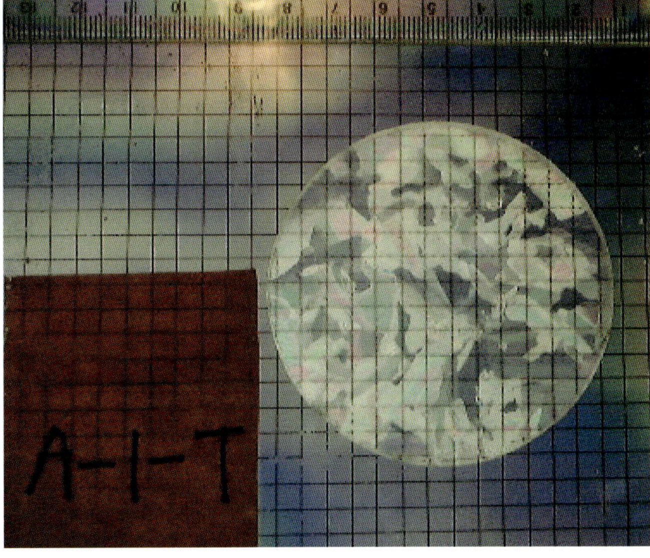

▲ Uniaxial compression test and a polarized photo of ice specimen

As natural resources development becomes active, cities are being constructed along the coast of the Arctic Ocean, and the Arctic sea route is not only used to transport goods to Arctic coastal cities, but is also recently used as the shortest trade route linking East Asia and Western Europe. An icebreaker is required to transport goods through the Arctic sea route. Icebreaker is a vessel that has the ability to independently navigate the ice-covered sea. Because Korea does not have direct contact with the polar environment, the academic research and technology level in the field of polar engineering was relatively inferior. It can be seen that the current level has been formed to some extent, and now is the time to gather them and effectively respond to the future demand.

▲ Preparation of model ice for towing test in the KRISO ice model basin (Courtesy of KRISO)

Diminishing Polar Sea Ice and Climate Change

In September 2012, Arctic sea ice extent shrank to 3.41 million square kilometers, the smallest since satellite observations began in 1979. This is a reduction of 50% compared to the summer minimum area between 1979 and 2000, and again breaks the record for the minimum area of 4.17 million square kilometers recorded in September 2007. It is predicted that it will be difficult to see summer sea ice in the Arctic before 2030. The shrinking of the sea ice extent means that the white space where sunlight is reflected is reduced, which raises the sea surface temperature in the polar regions and eventually accelerates the melting of sea ice.

The decrease in polar sea ice caused by global warming is already changing our lives in many ways. Since 2006, sea routes through the Arctic Ocean have been opened simultaneously along the continental coast, and in the past, the area was inaccessible due to thick sea ice. What was found in the 2010 Arctic

▾ Changes in Arctic sea ice extent (NSIDC/Climate Central)

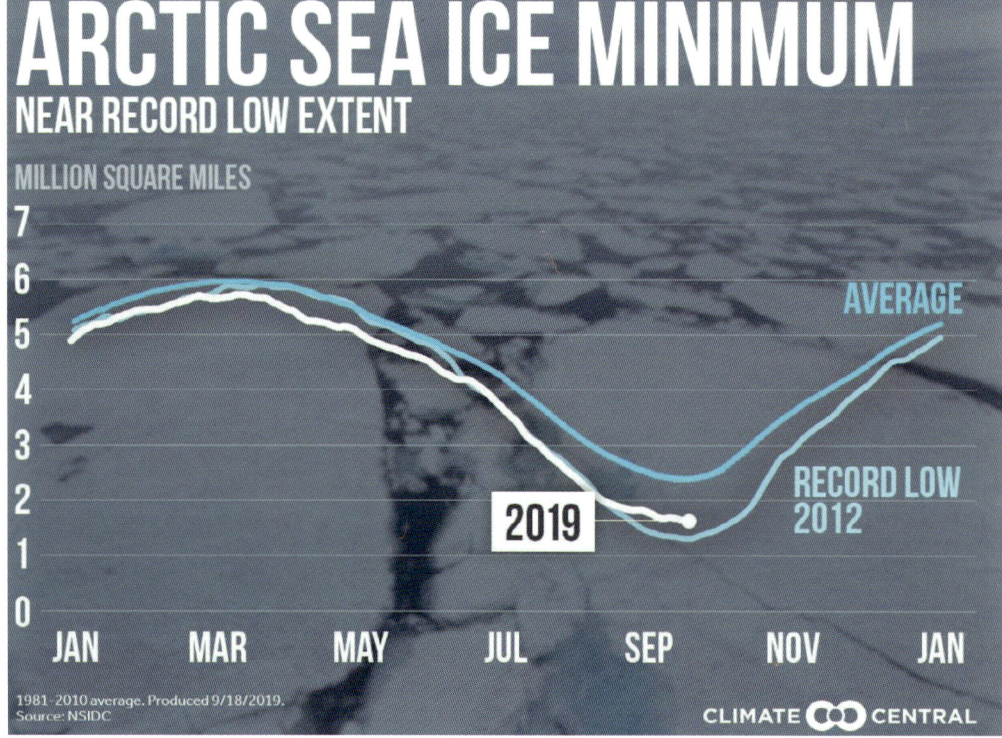

summer ice field test using the icebreaking research vessel ARAON was that a big ice sheet suited for the icebreaking test could be found at 78°N, but in 2012, it was only up to 82°N that a similar ice sheet was found.

The decrease in polar sea ice and dramatic change in climate have a negative impact on human life. The reduction of polar sea ice caused by global warming and its cascading effects increase the concentration of carbon dioxide in the atmosphere and may face the threat of ocean acidification and sea level rise. More directly, the reduction of polar sea ice destroys the ecosystems of animals and plants living in the polar regions and also causes abnormalities in ocean currents and atmospheric circulation, which increases the possibility of severe cold and strong storms, especially in the northern hemisphere. There is an urgent need for international efforts to understand the aspects of climate change detected in these bipolar regions at a global level and to find solutions.

▾ Arctic sea routes across the Arctic Ocean (NASA)

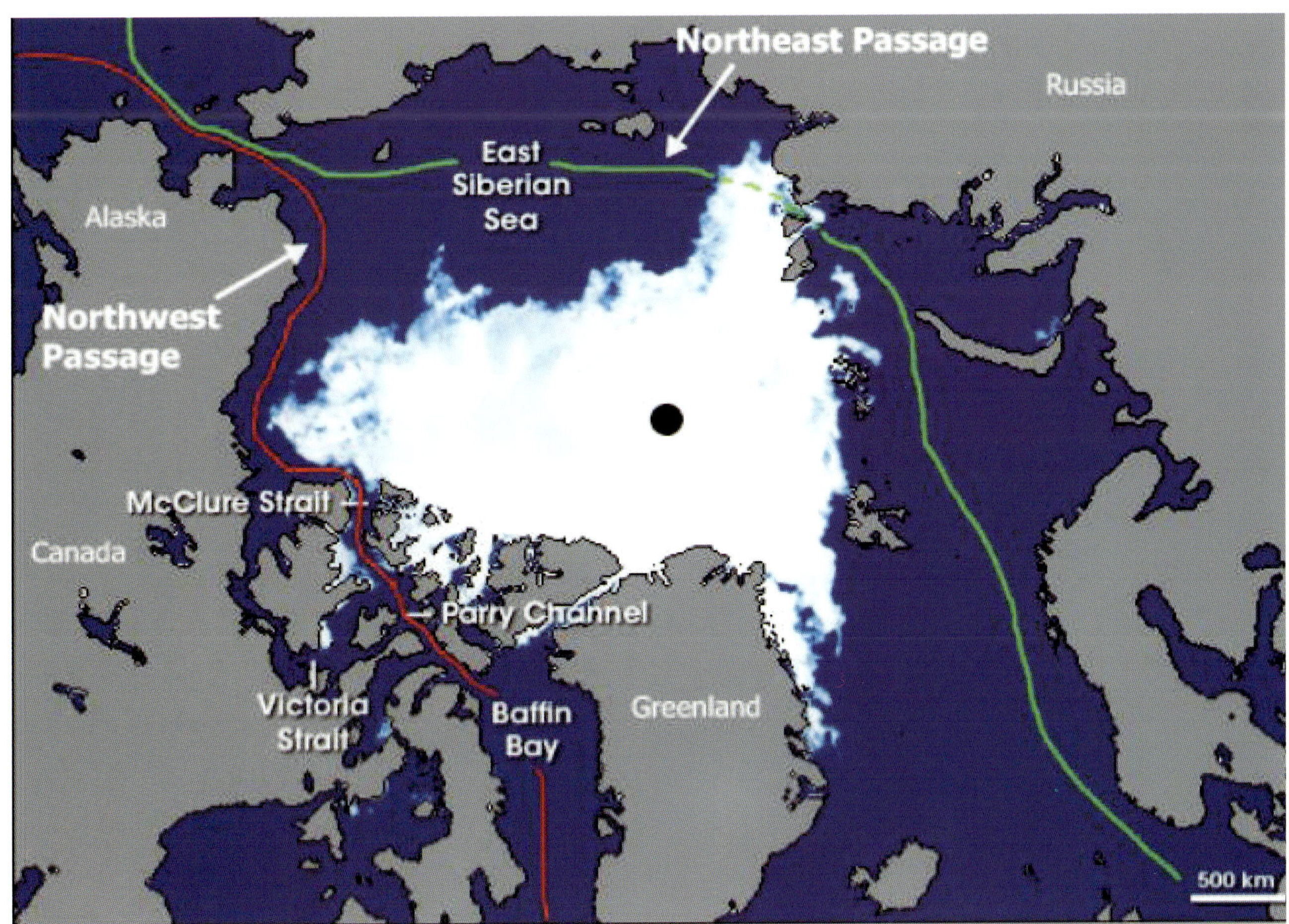

Formation and Development of Sea Ice

1) Freezing of fresh water and sea water

Ice that appears in the polar regions has various forms and there are regional differences, but basically, it can be divided into two types according to the source of its formation. First, sea ice is ice formed when sea water freezes, and salts in sea water are also contained in the ice. The other type is land-based ice, which is a glacier and an iceberg that has fallen into the sea, so this ice contains no salt. The World Meteorological Organization (WMO) has compiled and published terms related to sea ice used academically. According to the formation and development of sea ice, the shape and size of sea ice vary, and the name changes, and the structural strength of sea ice also changes accordingly.

Since ice is formed from liquid water, it retains some of the geometric shape of the water molecules. In liquid water, two hydrogen atoms are bonded to one oxygen atom at an angle of about 105°, and the overall molecular structure is approximately tetrahedral. The molecular structure of ice is based on the tetrahedral shape of a water molecule. The crystal structure of ice is a continuous tetrahedral arrangement of oxygen atoms, and each oxygen atom is surrounded by four other oxygen atoms and a hydrogen atom sandwiched between them. Oxygen atoms are tightly bound in symmetrical hexagonal layers, but relatively loosely bound between layers. This is the cause of the difference in structural strength depending on the direction and plays an important role in explaining the crystal structure of sea ice.

Fresh water has a maximum density at 4°C before

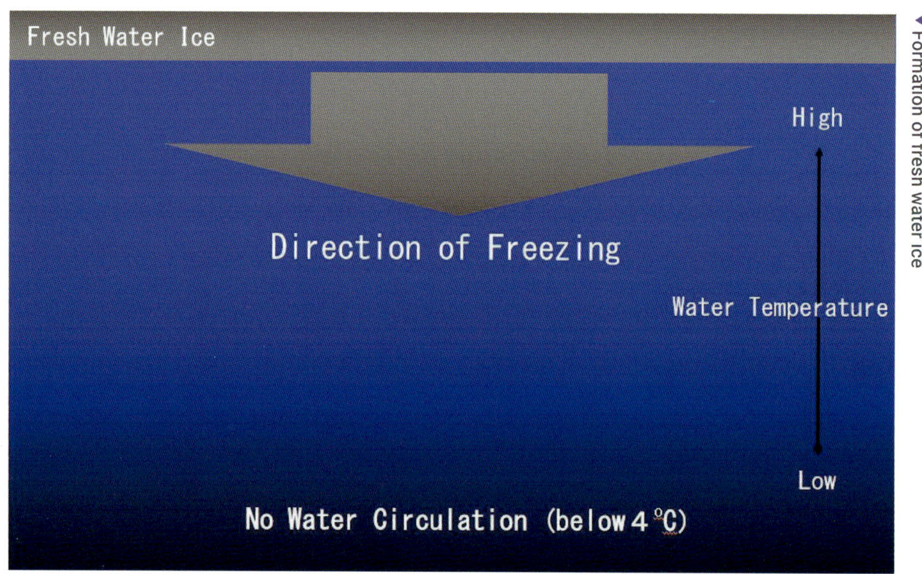

▲ Formation of fresh water ice

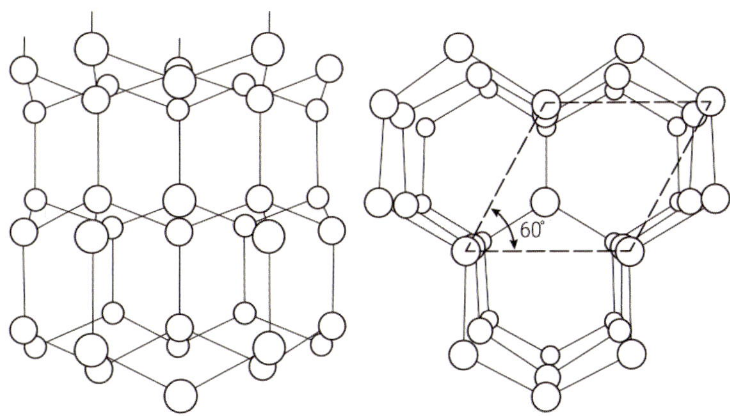

Hexagonal arrays of pure ice crystals (from Sanderson,1988)

freezing, whereas sea water gradually becomes heavier as the temperature approaches its freezing point. When the surface temperature of fresh water drops below 4°C and approaches 0°C, cold and light water collects on the surface and does not mix well with the water at 4°C below it. Once the surface temperature of fresh water drops below 0°C, the surface water begins to freeze very easily and soon forms a stable ice sheet.

Sea water, on the other hand, becomes heavier as the surface layer cools and sinks until it reaches equilibrium with the relatively heavy layer below it by convection. Warm, fresh (less salty) sea water is replenished instead of sunken cold sea water. If the salinity of the whole sea is uniform, the whole sea water will have to be mixed in this way until it reaches a uniform temperature of -1.7°C, after which the sea water will begin to freeze. In practice, however, the underlying sea water is generally more salty, which makes the bottom water heavier than the cooler surface water. Therefore, only the surface water of about 10~20 m in the upper layer cools close to the freezing point, and it does not mix well with the high salinity sea water in the rest of the lower layer. This is the reason why ice does not freeze well in the sea compared to lakes.

Sea water freezes at about -1.7°C, but the freezing temperature depends on the amount of salt it contains. The surface sea water of the Arctic Ocean in winter contains salinity about 3.0 to 3.4%. In summer, river water flows in or ice melts, so sea water is relatively warm and low in salinity.

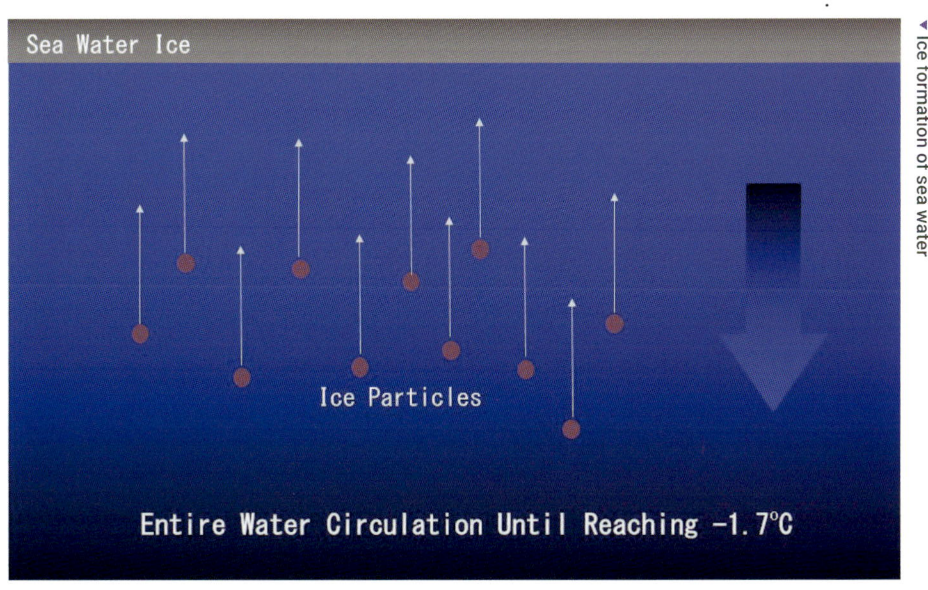

Ice formation of sea water

2) Development process of sea ice

When the seawater on the surface is sufficiently cold and the temperature is kept below -1.7°C, small crystals of ice (frazil ice) begin to form and float on the surface. This ice has a shiny surface like silver oil, so it is called grease ice.

The ice made in this way is broken by waves and collides with each other to form an irregularly round ice sheet with a thickness of about 10 cm. This is called pancake ice. The next step is to bond around the pancake ice to form young ice, a very stable and hard ice sheet. Once an ice layer is formed on the water surface, unlike before, the bonds between the ice pieces no longer occur, but continue to grow in the direction of the thickness of the ice sheet. Although it depends on the temperature, the increase in the thickness of the ice sheet is usually about 1 cm per day, and the ice crystals grow downwards of the ice sheet.

After that, the thickness of the ice sheet becomes thicker due to the cold temperature in winter and first-year ice is formed. In the summer of the following year, the temperature rises, and if the previously formed first-year ice does not completely melt during the summer, in the following winter, the first-year ice freezes again and becomes thicker. This ice that survives two winters is called second-year ice, and ice that survives several years of thawing and freezing is called multi-year ice. As the ice sheet thickens due to freezing, salt escapes from the contact surface of ice and seawater and is trapped between ice crystals. The amount of salt contained in sea ice is determined by the amount of salt trapped between ice crystals, and this is a big difference between first-year

▼ Frazil Ice

▼ Multi-year Ice

▼ Pancake Ice

▼ Melting Ponds

ice and multi-year ice. Multi-year ice has a strong texture and has a smoother surface than the first-year ice in appearance.

Every autumn, sea water begins to freeze from the periphery of the permafrost, and the entire Arctic Ocean is covered with ice from May to June of the following year. In winter, the average thickness is about 5-6 m, and in summer, the average thickness is about 2-3 m. Due to the rotation of the earth and the movement of ocean currents, the polar ice of the Arctic Ocean rotates slowly clockwise around 80°N and 150°W. These movements do not occur integrally, and it is known that the vicinity of the center rotates with a cycle of about 3 years, whereas it is known that the area rotates with a cycle of about 10 years in the outskirts. The movement speed of ice measured from the outskirts is about 1 to 2.5 km per day, but in the spring when the ice starts to melt, it moves as fast as 25 km per day in some cases.

During winter, a relatively uniformly thick ice sheet, called land fast Ice, is formed on the shoreline adjacent to the land, which grows to a maximum thickness of about 2 m in spring time and gradually melts and disappears in summer. On the other hand, the polar pack ice in the Arctic Ocean is a multi-year ice with balanced freezing and thawing, and is a huge mass of ice that has been split by intermediate open channels. Due to the movement of polar ice in the Arctic Ocean, a number of open channels are formed in the boundary zone between fixed ice and multi-year polar ice, allowing the passage of ships even for a short period of time.

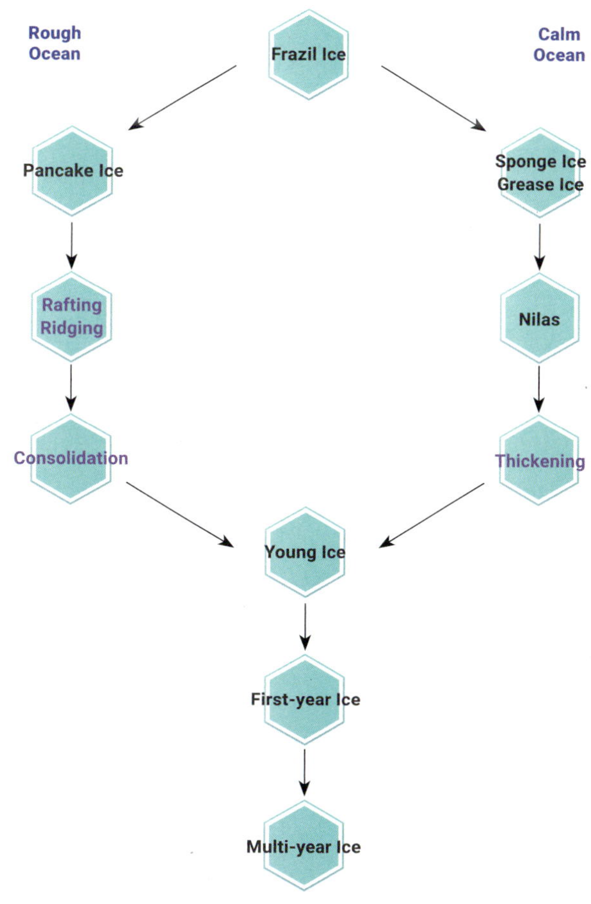

Sea ice development process from ice crystals to sheet ice

Research Sites of Polar Sea Ice

1) Kemi, Finland and Svalbard, Norway

Prior to the full-scale Antarctic sea ice research using the icebreaker research vessel ARAON in accordance with the research project, in April 2010, Kemi, Finland and Svalbard, Norway were visited to practice measuring instruments and related techniques. The northern Baltic Sea is a closed sea area surrounded by Sweden, Finland. In Kemi, Finland, located at the tip of the Gulf of Bosnia, 80 cm thick sea ice was formed even in April, so several icebreakers were working to open the navigational routes. The research team boarded the icebreaker and had the opportunity to investigate the sea ice in the Baltic Sea. SAMPO, an icebreaker with a length of 75 m and a displacement of 3,500 tons, was put into service for 30 years to open a winter route in the Baltic Sea, but is now retired.

Norway's Svalbard Archipelago lies at the mouth of the Arctic Ocean, 650 km from the northern tip of Europe, and extends from latitudes 74°N to 81°N. It has a higher latitude than Iceland, Siberia or Alaska. The average temperature in winter is about -15°C and the temperature rises to zero for about four months in summer. Although the Svalbard Islands are Norwegian territory, they are recognized as the international commons by the Svalbard Treaty. The largest island Is Spitzbergen, and in many places coal mines have been developed and operated for a long time. In Ny-Ålesund, Svalbard, Korea's Dasan Science Station, is located,

▾ Research sites of sea ice study - Kemi, Finland and Svalbard, Norway

and in the capital city, Longyearbyen, UNIS (University Center in Svalbard), an international cooperation university, is located. Various studies using the environment are being conducted. The 'Noah's Ark' project, which provides a seed storage facility in the Arctic permafrost, was also installed in an abandoned mine in Longyearbyen. Our research team was also able to receive training on sea ice measurement and how to use measuring instruments with the help of UNIS. The research team's sea ice study was carried out in Svea Bay for two days in early April 2010 by snowmobiles to Svea, a coal mining village 60 km away from Longyearbyen.

▼ Finnish icebreaker SAMPO at the port of Kemi

▼ Researchers on board the Finnish icebreaker SAMPO in the Baltic Sea ice

Sea ice research in Svea, a mining town 60km away from Longyearbyen ▸

▾▴ Houses in Longyearbyen and the abandoned coal mine facility

UNIS (University Center in Svalbard) located in Longyearbyen

UNIS (University Center in Svalbard) and snowmobiles

▼ Global seed vault located in Longyearbyen, Svalbard

Equipment packed on the snow sleigh before leaving for Svea, sea ice research site

2) Chukchi Sea and Beaufort Sea in the Arctic

Located north of 75°N, the Central Arctic Ocean is covered with polar pack ice. It's not the same ice every year, nor is the ice sheet completely made up of one mass. Arctic sea ice is in dynamic equilibrium, where fragments of multi-year ice break off from the outer boundary of the polar pack ice and flow into the southern seas. When a crack occurs inside the polar pack ice, the newly exposed open channel immediately freezes and repeats the process.

The Chukchi Sea and the Beaufort Sea are located south of the Central Arctic Ocean, adjacent to the coasts of Alaska and Canada. Usually, freezing starts in October and thawing starts in June of the following year, but the period of freezing is shorter than in the past due to the recent global warming effect. This sea area is covered with ice sheets with an ice concentration of 10/10 in the middle of winter, and even when the thawing season is near, the density is usually more than 5/10. A density of 5/10 corresponds to a condition in which it is difficult for ordinary vessels other than icebreakers to navigate. In summer, about 5-30% of the surface area of the Beaufort Sea is covered with multi-year ice, and the closer to the shore, the more the ice disappears and the sea becomes ice free.

The period of investigation of sea ice in the Chukchi Sea and Beaufort Sea aboard the icebreaking research vessel ARAON was about 30 days in July and August 2010. Three icebreaking performance tests were carried out and the material properties of sea ice were investigated on the ice sheet. The test site is a place where there is a big ice sheet that can sustain icebreaking ability of the ARAON, took place at 78°N. In summer 2015, another sea ice field test was carried out in the Beaufort Sea.

▾ Research sites of sea ice study - Chukchi Sea and Beaufort Sea in the Arctic

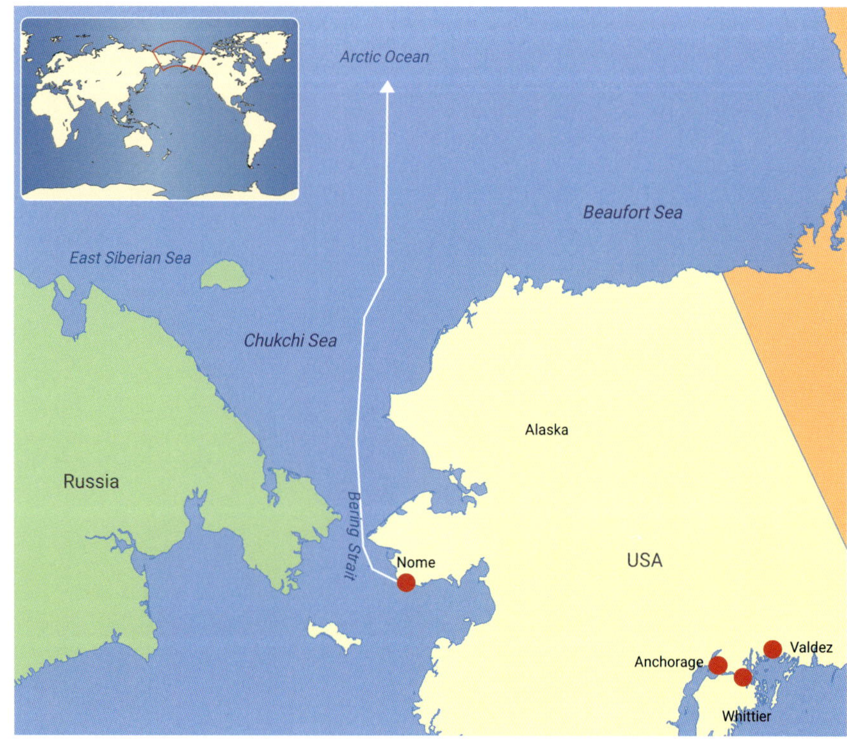

▲ Research team preparing for equipment and waiting for disembarkation

▼ Sea ice research team for the Arctic Sea Expedition 2010

▼ Houses in Nome,
Alaska and the ARAON seen from the airplane

3) Amundsen Sea and Ross Sea, Antarctica

The ice sheets formed in the Antarctic Ocean have a greater seasonal variation than in the Arctic Ocean. The sea ice around Antarctica reaches a maximum area of 20 million square kilometers in September and shrinks to a minimum area of 13 million square kilometers in March. The reason for this large seasonal variation is that Antarctica is a continent surrounded by the sea, and the perimeter of the sea ice is directly exposed to the sea, so ice can freely spread out from the polar pack ice in winter. The Antarctic convergence also plays an important role in determining the maximum extent of sea ice. The convergence line refers to a sea area where the temperature of surface water changes rapidly in the ocean, and is usually distinguished when a change of 5°C occurs at a distance of 250 km. The Antarctic convergence line exists between 47°S and 63°S and its location is closely related to the maximum extent of sea ice.

In general, the ice in the Antarctic Ocean is flat and less curved because it is subjected to less dynamic action than the ice in the Arctic Ocean. In the Antarctic Ocean, ice spreads toward the outskirts unconstrained by the action of wind and ocean currents, and the possibility of mutual collision is slim. About 15% of sea ice in Antarctica can be considered multi-year ice, especially concentrated in the west of the Weddell Sea, in the Amundsen Sea and in the Bellingshausen Sea.

The most distinctive feature of the Antarctic sea ice is the huge icebergs that broke off from the Antarctic ice cap and the surrounding glaciers. Icebergs in the Antarctic Ocean are usually 500 m in diameter and 250-450 m thick, but giant icebergs exceeding 300 × 100 km have also been observed. Icebergs drift at a rate of 15 to 20 km per day, depending on wind and currents.

▾ ARAON at Lyttelton port, Christchurch, New Zealand

Research sites of sea ice study - Amundsen Sea and Ross Sea

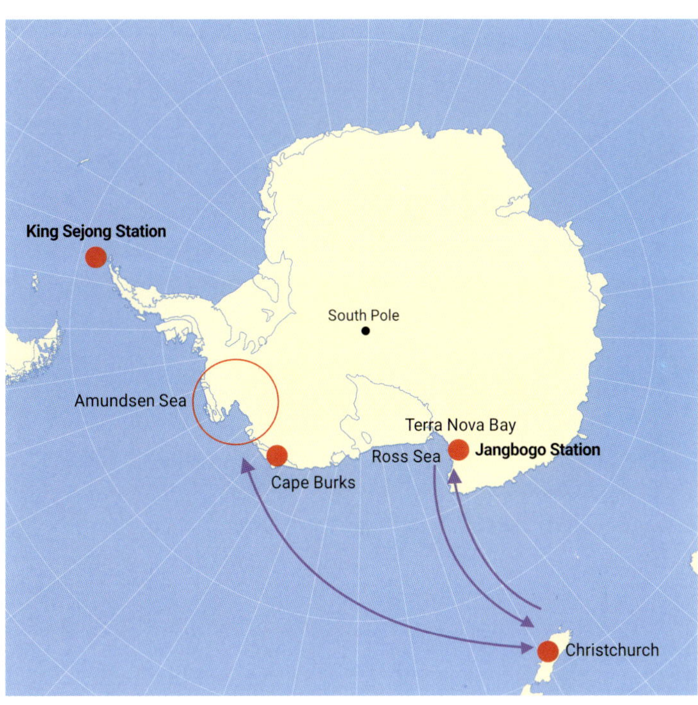

Lyttelton port seen from the ARAON

▲ BBQ party on sea ice after measuring ice work in the Amundsen Sea

Amundsen Sea is located in the western part of Antarctica, and the recent glacial loss in Antarctica has also been prominent, providing a major factor in global warming, and it is a place where changes in the ocean circulation around Antarctica can be clearly detected. In February and March 2012, for about 55 days in February and March 2012, the icebreaker research vessel ARAON departed from Christchurch, New Zealand, and visited the Amundsen Sea and surrounding waters. The sea ice study area of the Amundsen Sea was between 72°S and 75°S, where there is a lot of multi-year sea ice. In December 2019, the ARAON conducted an ice field test in Ross Sea, Antarctica, where the second Korean science base, Jangbogo Station, is located nearby. During the field test, extensive amount of data was recorded from strain gauges also from fiber optic sensors installed on the inside hull plate of ARAON's bow area. Information such as ship speed, ice properties and ice thickness were also collected from various sensors and integrated monitoring system.

Morning sunlight on the way to Lyttelton harbor after 55-days Antarctic expedition ▲

Part 1.
Early Stages of Sea Ice Formation

Frazil Ice

Frazil ice refers to needle-shaped ice crystals with a diameter of 3 to 4 mm that are formed in sea water cooled below the freezing point as the first stage of sea ice formation. Frazil ice is usually formed near the surface of turbulent rivers, lakes and oceans on cold days when the atmospheric temperature drops below -6°C.

1 3 Amundsen Sea

2 4 Beaufort Sea

1

2

3

4

Frazil Ice

1 ~ 4 Amundsen Sea

1

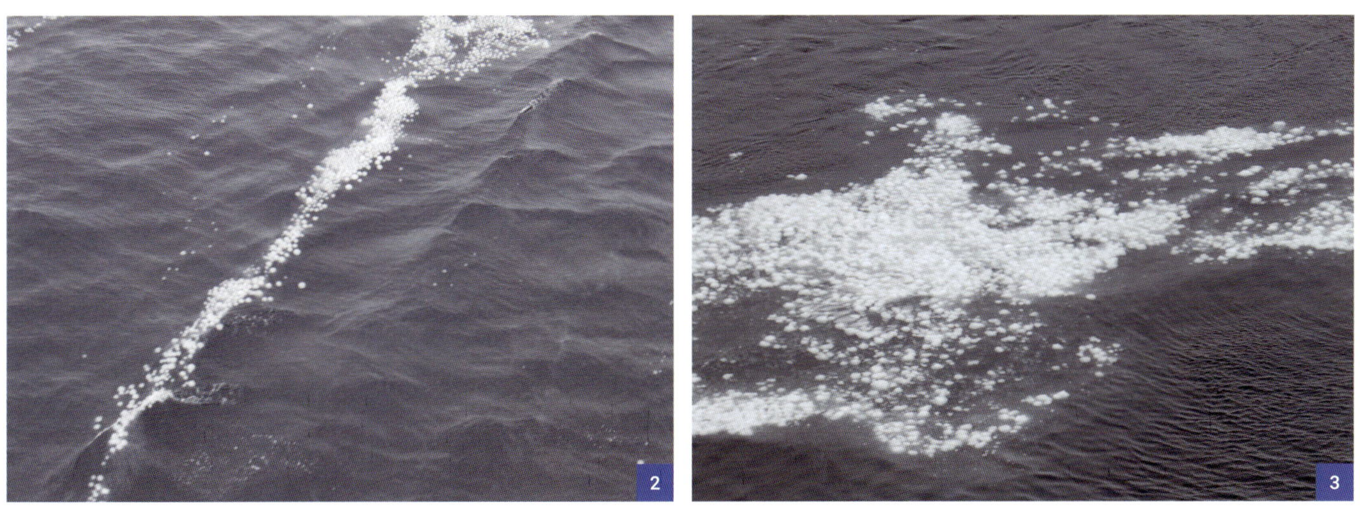

Mixture of frazil ice and old ice found in the Amundsen Sea ▼

Sponge Ice

Frazil ice forms a thick layer like a sponge or ice cream slush in sea water.

1 ~ 4 Amundsen Sea Polynya

1

2

3

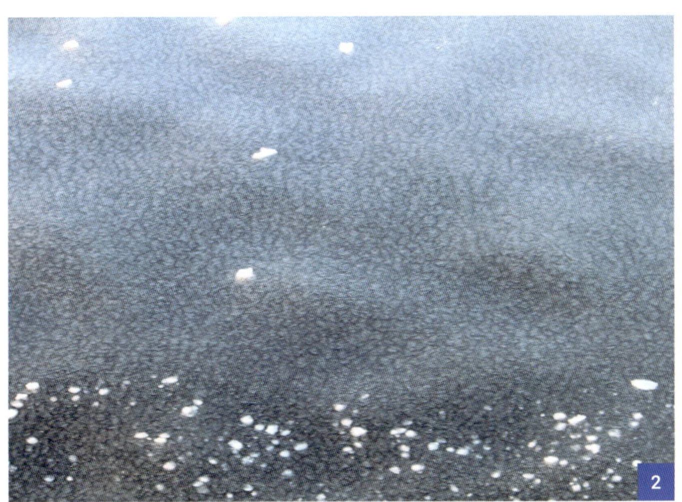

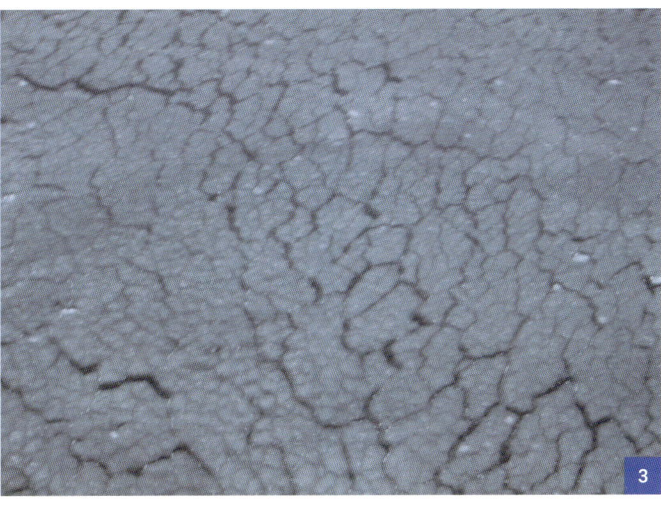

Sponge ice found near Cape Burks, Amundsen Sea ▼

4

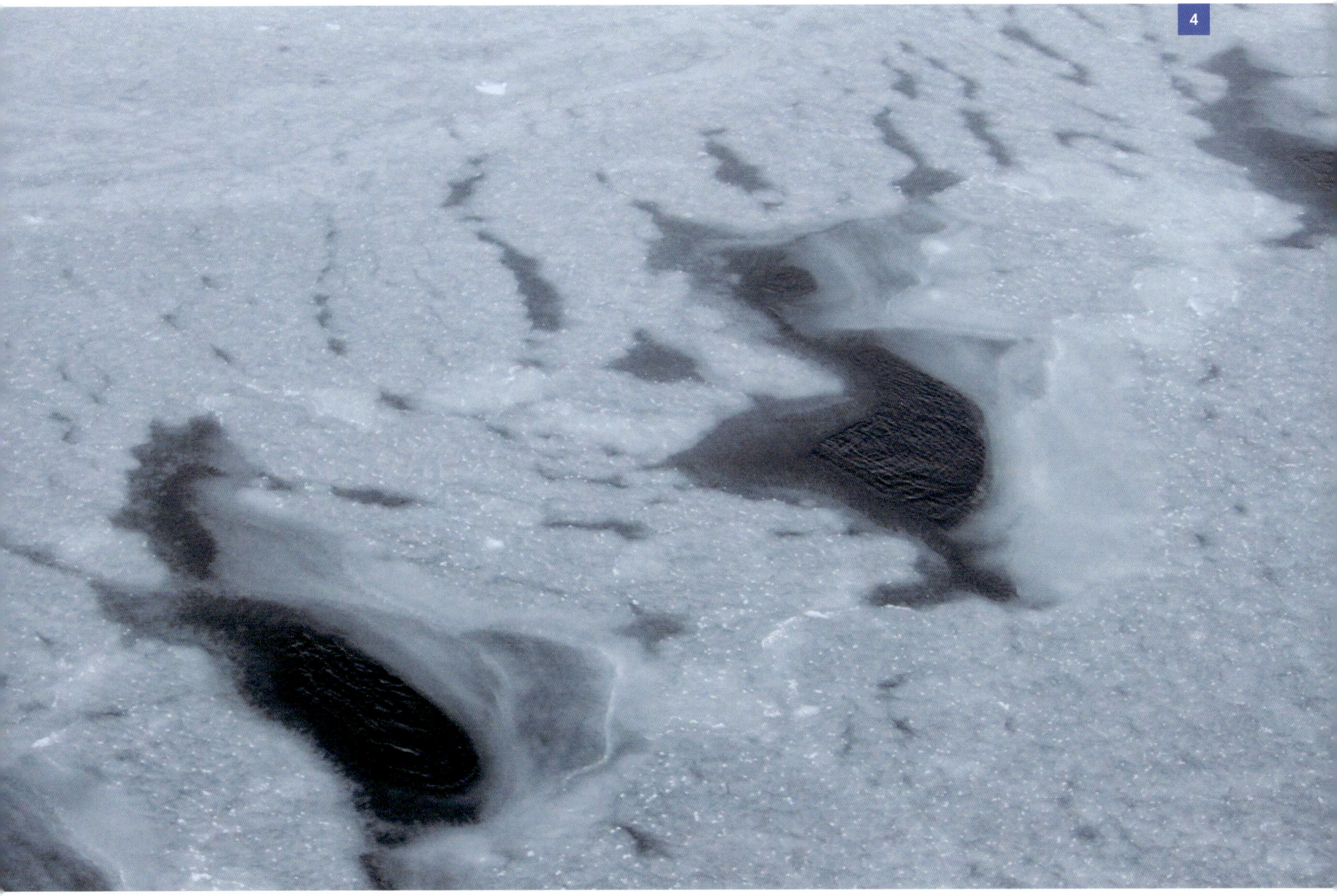

Grease Ice

The second freezing stage after frazil ice, where ice crystals gather at a relatively calm water surface without sea level fluctuations to form a thin, smooth layer of ice. This layer of ice doesn't reflect much light, creating a shiny but dark surface that looks like an oil spill.

1 ~ 4 Amundsen Sea

1

4

Grease Ice

1 4 Longyearbyen, Svalbard

2 3 Amundsen Sea

▼ Thin ice crust formed near Longyearbyen, Svalbard

1

2

Cose-ups of greasa ice ▲

3

Channel made in the grease ice ▲

Thin ice crust ▼

4

Nilas

Nilas is a more advanced state of grease ice, and is a thin, elastic layer of ice less than 10 cm thick. It is formed mainly on calm sea level and is thin and dark at the beginning of formation, but becomes lighter as it thickens. This relatively thin ice sheet is easily broken and overlapped by waves or wind, and finally the ice sheet grows stably and gradually thickens.

1 ~ 2 Newly frozen thin ice among old ice
3 Broken pieces of thin ice plate, Beaufort Sea
4 Thin ice deformed by waves in the Amundsen Sea

1

4

Rafting Ice

Because nilas are thin, they are easily broken by waves or wind and goes through the process of overlapping, giving the appearance of overlapping fingers like interlaced fingers (finger rafting ice). The ice crystals grow below the water surface and the ice sheet gradually thickens. Usually, a thin ice sheet with a thickness of about 30 cm before growing into first-year ice is called young ice.

1 4 Amundsen Sea

2 3 Beaufort Sea

1

2

3

4

Rafting Ice

1 ~ 4 Amundsen Sea

1

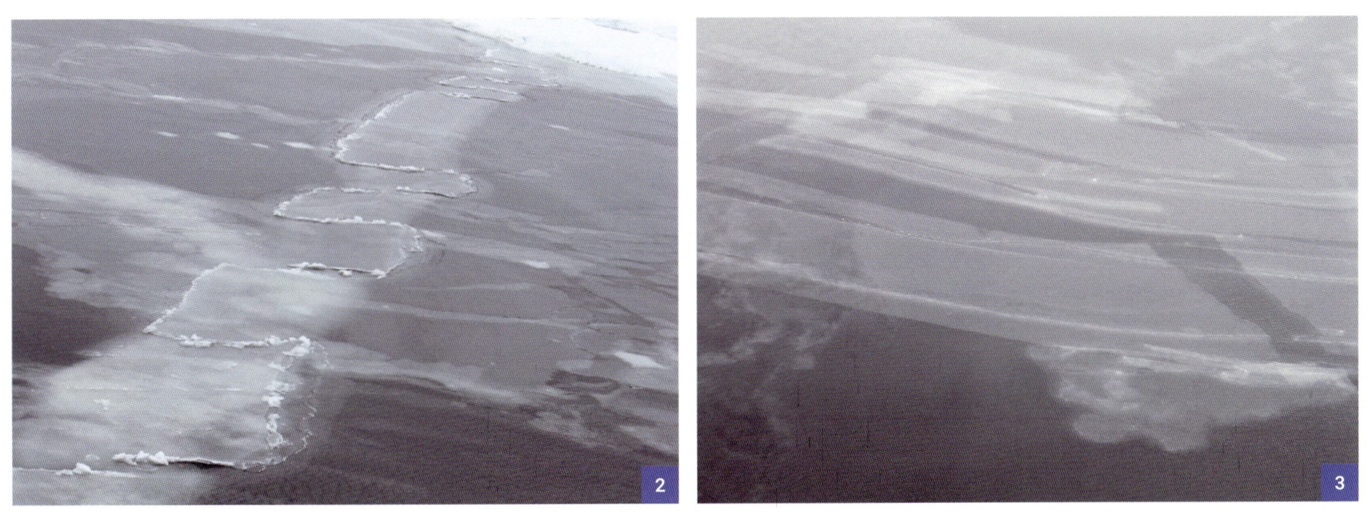

4

Rafting Ice

1 ~ 4 Amundsen Sea

▼ Newly developed young ice adjacent to old ice in the Amundsen Sea

1

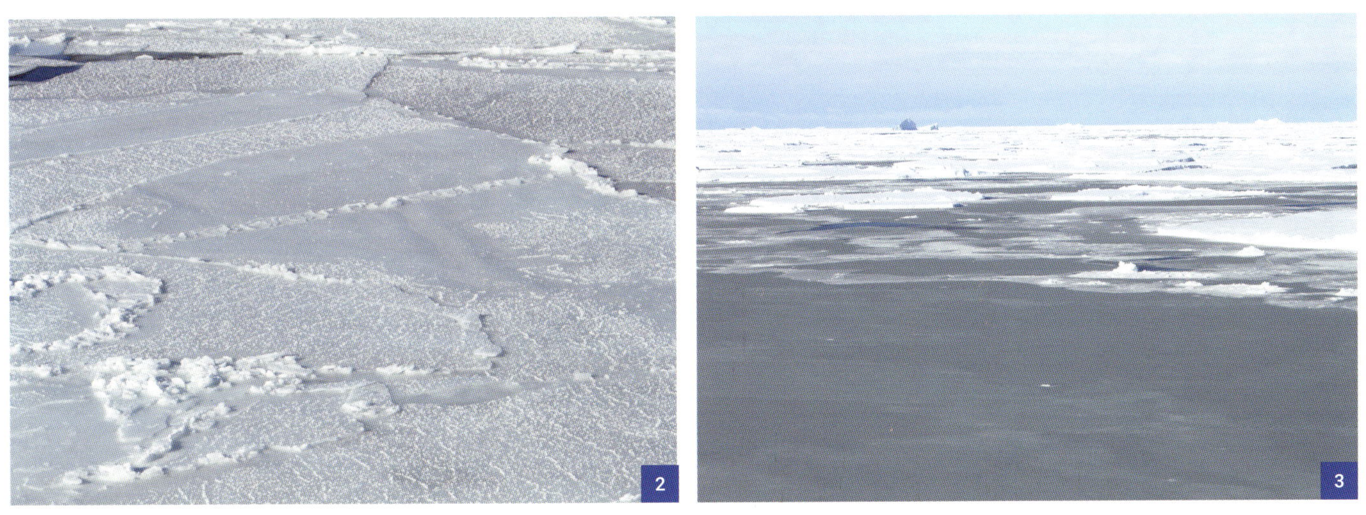

Pancake Ice
Initial Stage

When the sea surface is rough due to strong winds and waves, the already formed thin ice sheets are broken and collided with each other to form pancake ice with thick edges. Pancake ice is about 30 cm to 3 m in diameter and about 10 cm thick. Frazil ice or sponge ice crystals are agglomerated into a disk shape. Pancake ice, which was initially a loose cluster, grows into a thick ice sheet over a large area as the edges freeze together in the cold temperatures over the winter.

1 ~ 4 Amundsen Sea

3 Pancake ice formation in the Amundsen Sea (photo. by H.S. Kim)

▼ Pancake ice formation and morning light in the Amundsen Sea

1

Pancake ice with a reflection of the moon ▼

4

Pancake Ice

Intermediate Stage

1 ~ 4 Amundsen Sea Polynya

▾ Loose concentration of pancake ice in the Amundsen Sea Polynya

1

4

Pancake Ice
Intermediate Stage

1 ~ 4 Amundsen Sea

1

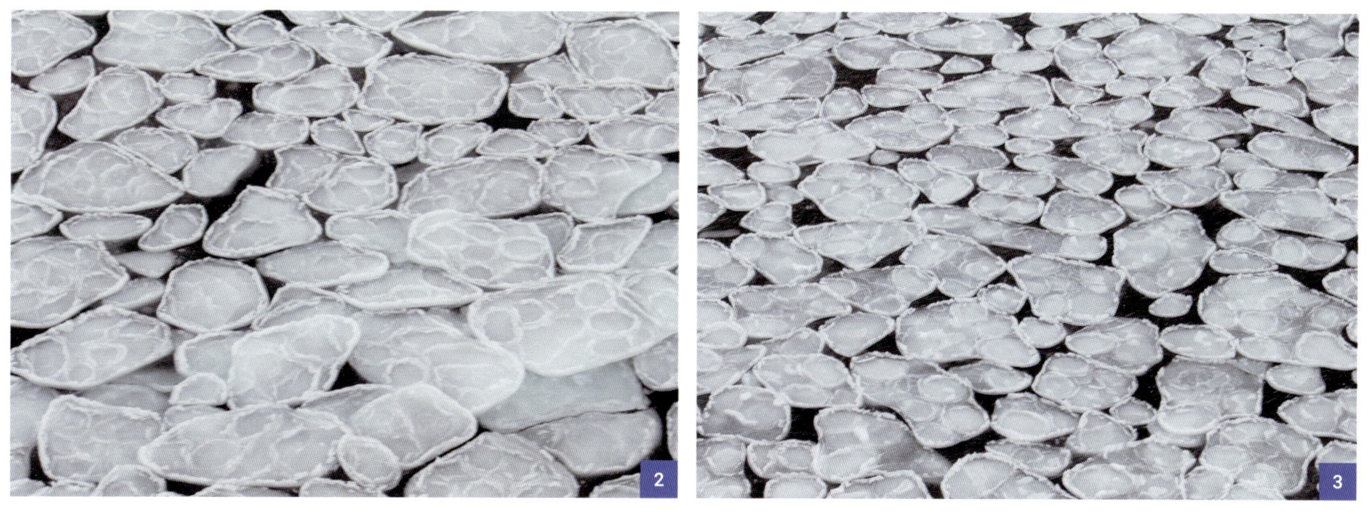

4

Pancake Ice
Final Stage

1 ~ 4 Amundsen Sea

▼ Dense concentration of pancake ice and formation of continuous ice plate in the Amundsen Sea

1

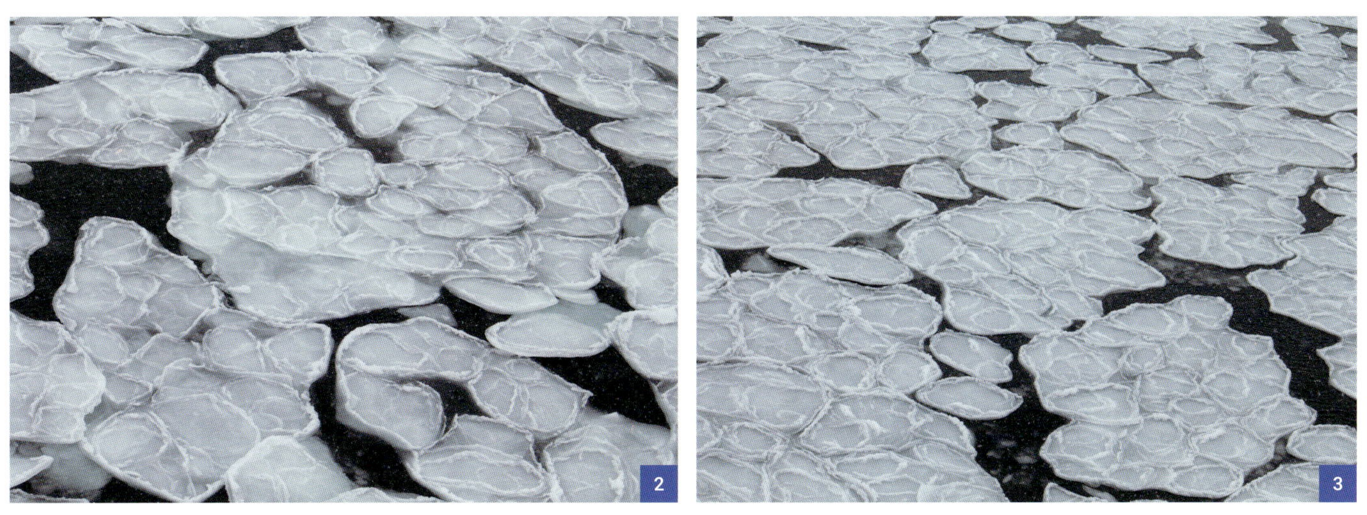

Pancake Ice

Final Stage

1 ~ 3 Amundsen Sea

4 Ross Sea

1

2

3

Quickly frozen pancake ice (Ross Sea) ▼

4

Ice in Waves

Waves do not occur in the sea where the surface is completely covered with thick sea ice, but there are waves in the periphery where polar ice meets the open sea, thin ice sheets, or where only a part of the water surface is covered with ice (usually ice concentration 7/10 or less), and the ice moves. In the early stages of pancake ice formation, the influence of waves is very large.

1 Southern Ocean

2 ~ 4 Amundsen Sea

▾ An iceberg drifting in waves

1

Pancake ice in waves ▲

Grease ice in the wake of IBRV ARAON ▲

Thin ice crust in waves ▼

Ice in Waves

1 Cape Burks, Amundsen Sea

2 3 Beaufort Sea

4 Southern Ocean

▼ Sponge ice in waves

1

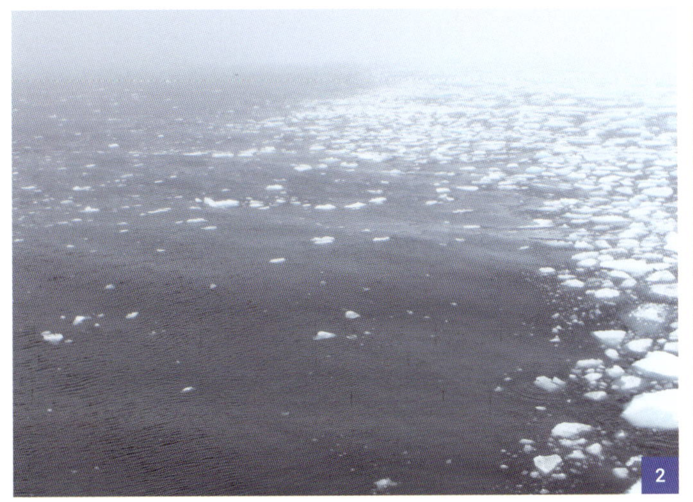

2

Boundary of ice field ▲

3

Boundary of ice field ▲

Lonely iceberg in rough seas ▼

4

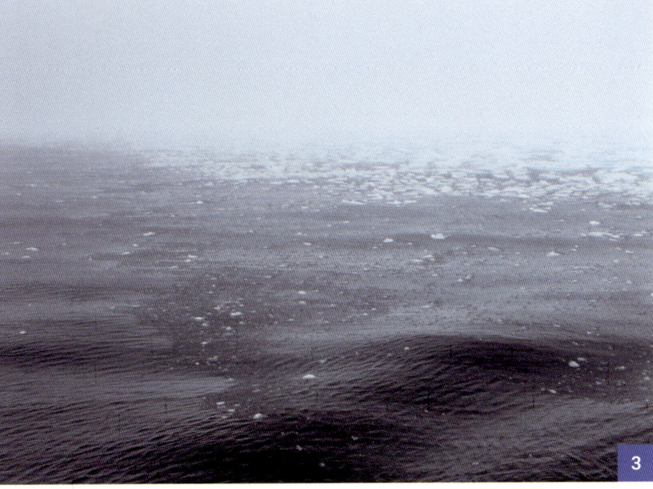

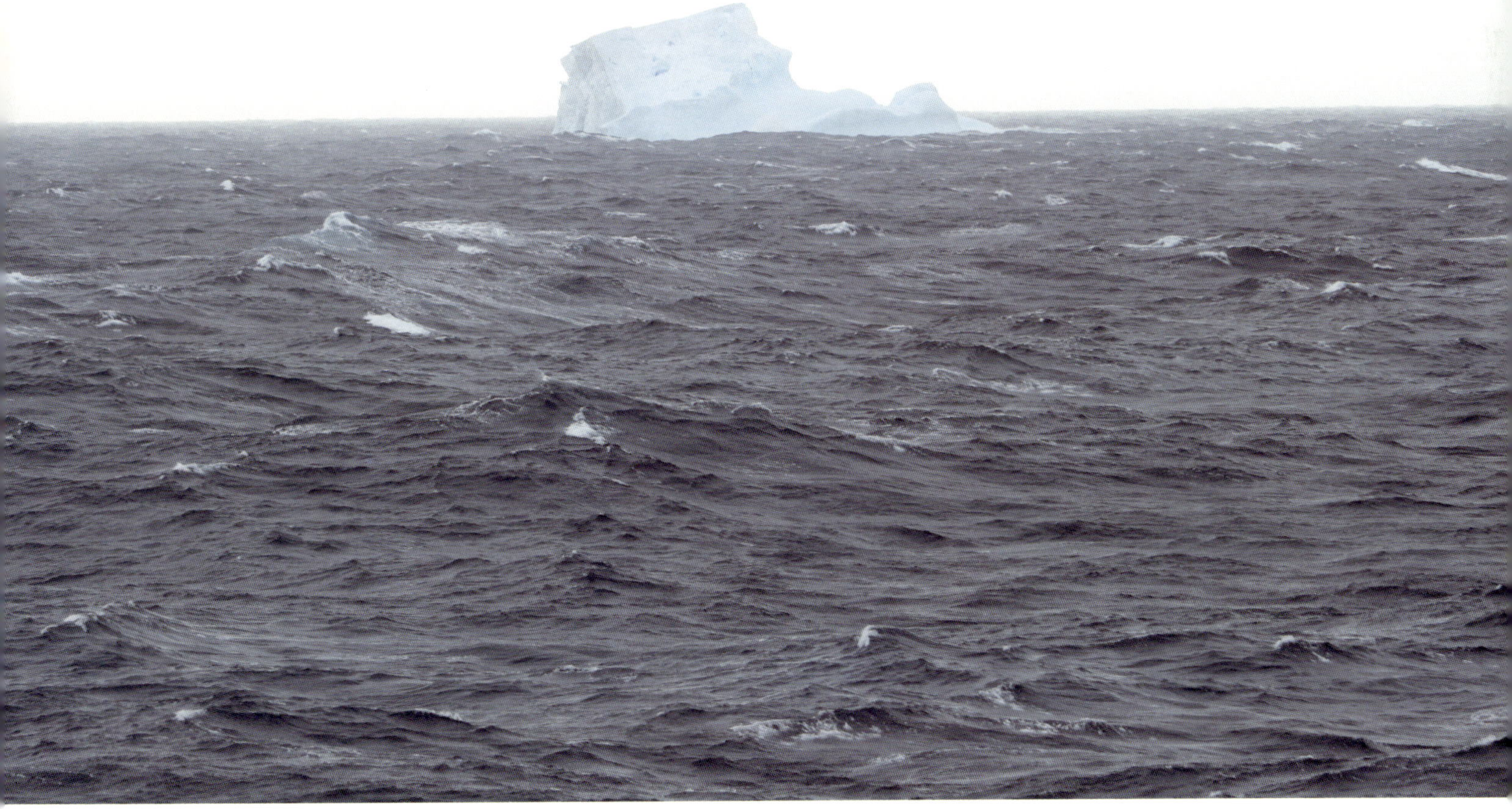

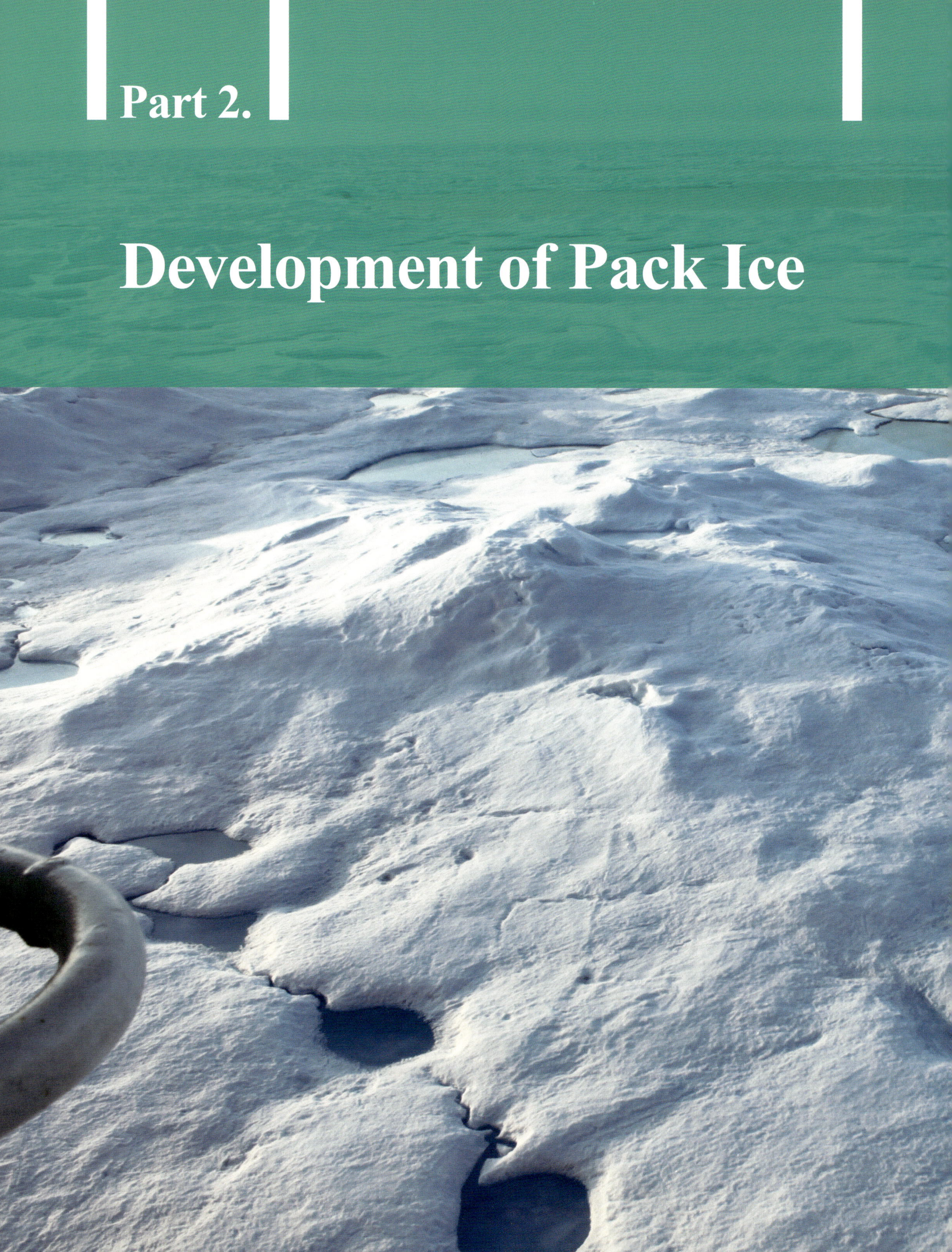

Part 2.

Development of Pack Ice

Land Fast Ice

Ice formed by adhering to the shoreline, between islands or in the sea deep into the land is called land fast ice. It is formed widely between icebergs attached to the wall of an iceberg or drift ice or stranded on a shallow shoreline. It is generally a first-year ice, and there are many types of level ice with a uniform thickness, and the width of land fast ice varies from several tens of meters from the shoreline spans hundreds of kilometers. On the other hand, pack ice refers to all types of ice floating other than fixed ice.

1 ~ 3 Svea, Svalbard

4 Ross Sea, Antarctica

▾ Land fast ice developed in the narrow inlet of Svea, Svalbard

1

2

3

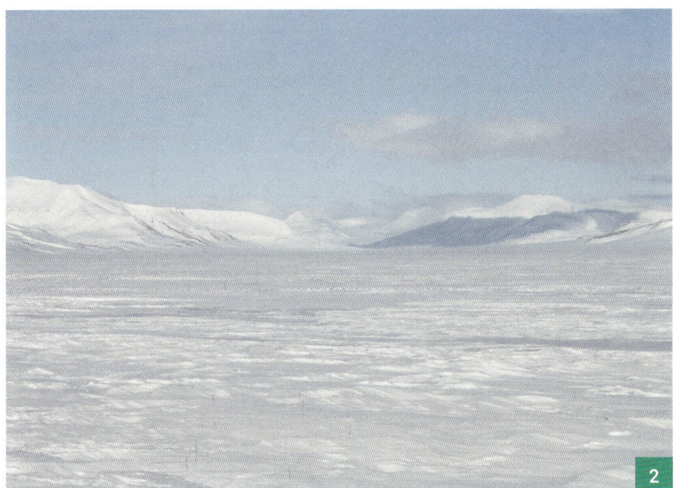

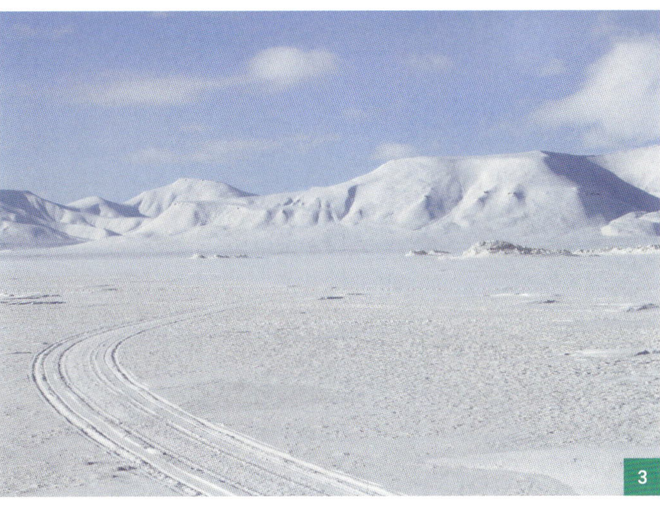

Land fast ice developed near Jangbogo Station, Ross Sea, Antarctica ▼

4

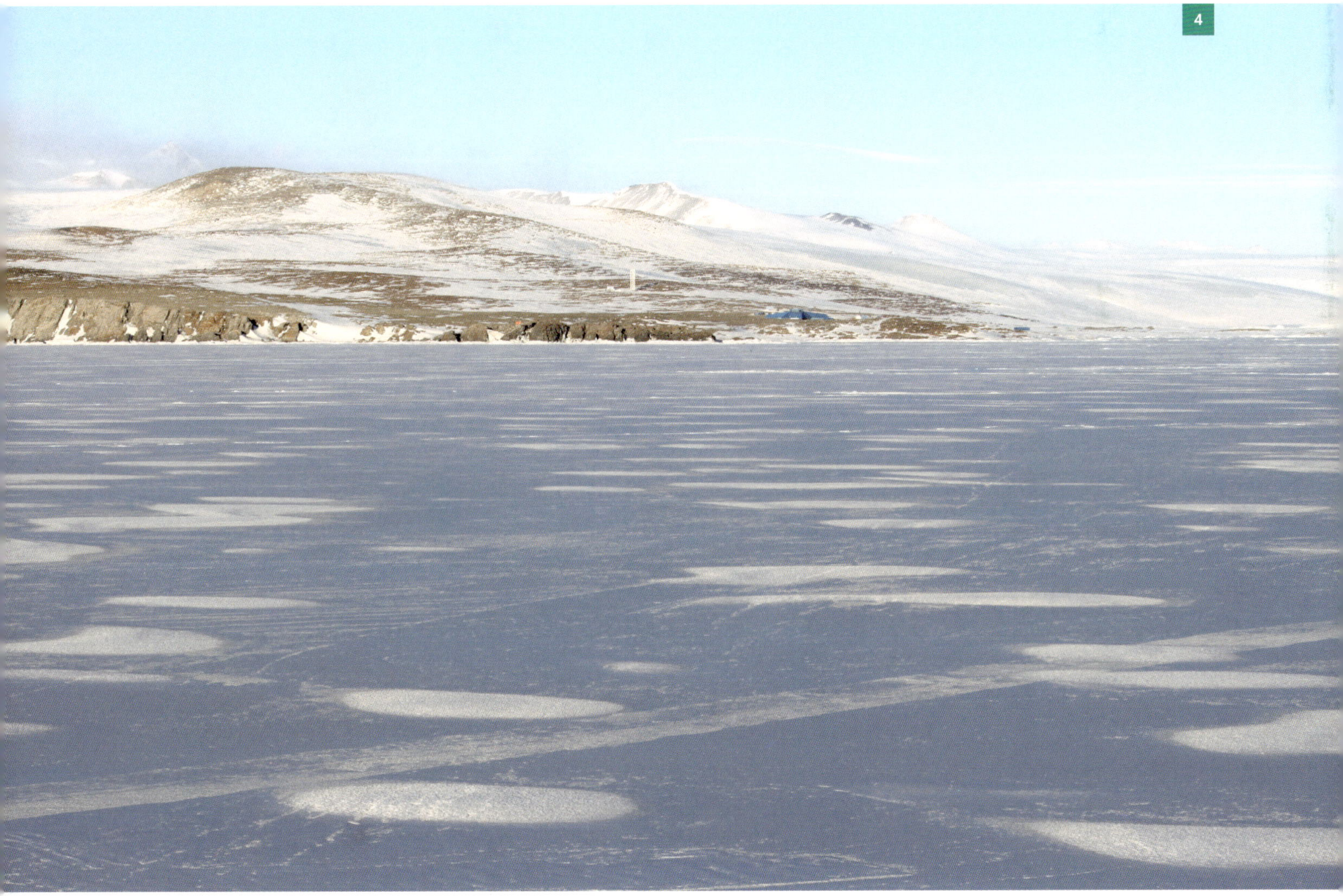

Level Ice and Ice Floe

Level Ice

It refers to flat sea ice of relatively uniform thickness before the ice sheets are rafted or changed to a ridged state like hills.

1 ~ 3 Amundsen Sea

▼ A foot print of the IBRV ARAON on level ice of 2m thickness

1

2

Thick level ice ▲

Moon rise over pack ice ▼

3

Level Ice and Ice Floe

Level Ice

1 2 Amundsen Sea

3 4 Northern Baltic Sea

▼ Multi-year level ice

1

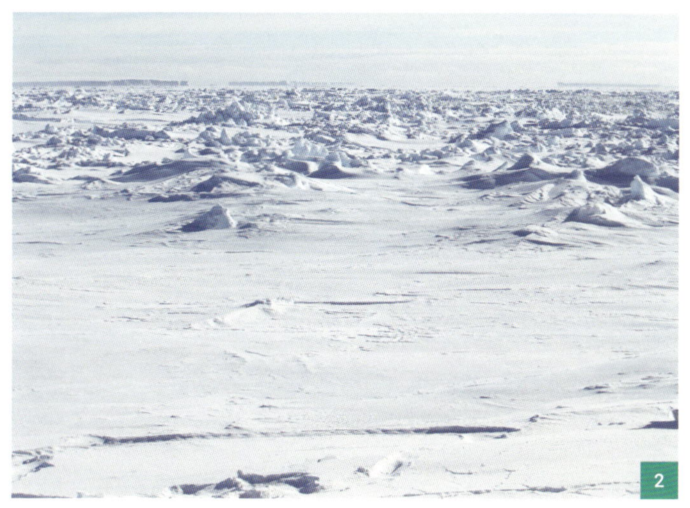

2
Multi-year level ice ▲

3
First-year level ice ▲

First-year level ice ▼
4
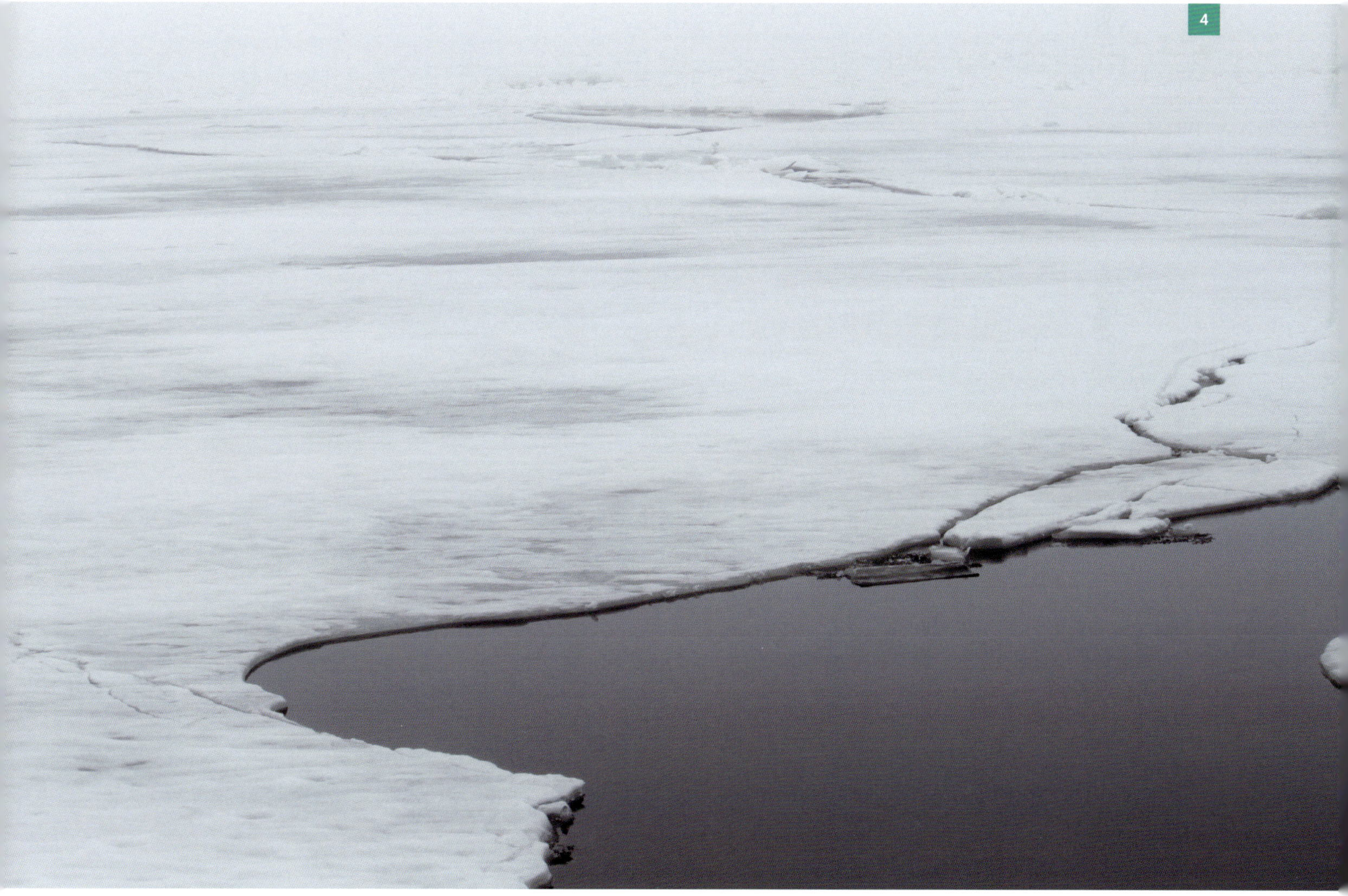

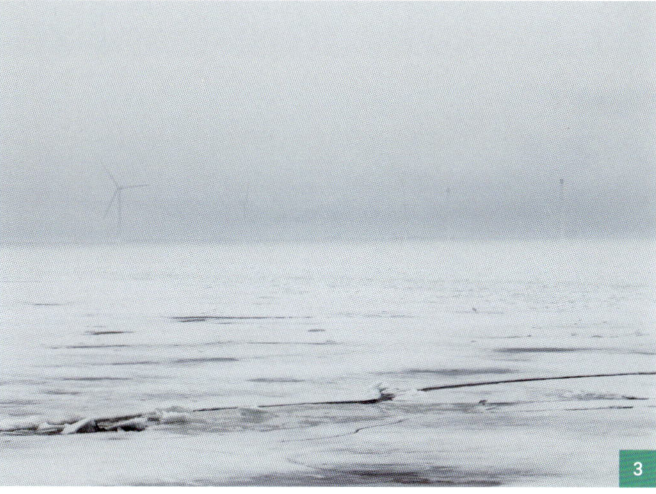

Level Ice and Ice Floe

Ice Floe

It is a relatively flat plate-shaped sea ice with a diameter of 20 m or more. Depending on the area, ice floes come in various sizes from huge ice sheets with a diameter of 10 km or more to small ice sheets with a diameter of 20 to 100 m.

1 ~ 4 Amundsen Sea

▾ Amundsen Sea covered with thick ice floe field

1

2

3

Pack ice filled with thick ice floes ▼

4

Level Ice and Ice Floe
Ice Floe

1 ~ 4 Amundsen Sea

▾ An iceberg among among ice floes

1

2

3

Consolidated thick ice floes ▼

4

First-year Ice

First-year ice is developed from young ice and has not passed more than one winter, and is about 30 cm to 2 m thick.

1 Northern Baltic Sea

2 ~ 4 Amundsen Sea

▾ First-year level ice developed near Kemi, northern Baltic Sea

1

2

3

Level ice selected for icebreaking test

IBRV ARAON's icebreaking performance test in the first-year level ice near Tera Nova Bay, Ross Sea (Courtesy of KOPRI) ▾

4

Multi-year Ice

Second-year ice is old ice that does not melt after only one summer, and is thicker and smoother than first-year ice. The one-time melting in the summer creates numerous irregular and small puddles. Areas where there is no snow or puddles are usually turquoise color. Multi-year ice is ice that has survived at least two summers without melting and is 3 m thick or thicker. Multi-year ice is smoother and less salty than second-year ice. When there is no snow, it usually appears blue.

1 2 3 Beaufort Sea

4 Amundsen Sea

▾ Multi-year ice field in the Beaufort Sea

1

2

3

Hummock on the multi-year ice ▼

4

Multi-year Ice

1 ~ 4 Amundsen Sea

1

2

3

4

Ice Ridges

The ice sheet is cracked and broken by wind and current and the broken ice pieces overlap each other to form a thick ice ridge. The surface is smooth and hard because multi-year ice ridges have been exposed to weathering for a longer period of time than first-year ice ridges. The multi-year ice ridge is also called hummock. The upper part of the ice ridge is called the sail and the lower part is called the keel. It is also observed to reach a height of 40 m. The existence of ice ridge is difficult to detect with the naked eye and becomes the biggest risk factor for the navigation of icebreakers.

1 ~ 3 Beaufort Sea
4 Baltic Sea

▾ Ice ridges formed in multi-year ice field

1

2

3

First-year ice ridge ▼

4

Ice Ridges

1 Beaufort Sea

2 ~ 4 Amundsen Sea

▼ Multi-year ice ridges

1

2

3

Multi-year ice ridges ▼

4

Melting Ponds on Sea Ice

Ice older than two years old melts on the top of the ice sheet during the summer thaw, forming numerous irregular, small pond-like puddles on the ice sheet. It is mainly caused by the melting of snow on the ice, but the melting of the ice itself in the next stage also contributes.

1 ~ 4 Beaufort Sea

▾ Melting ponds on the surface of old sea ice in the Arctic

1

Melting Ponds on Sea Ice

1 ~ 4 Beaufort Sea

1

2

3

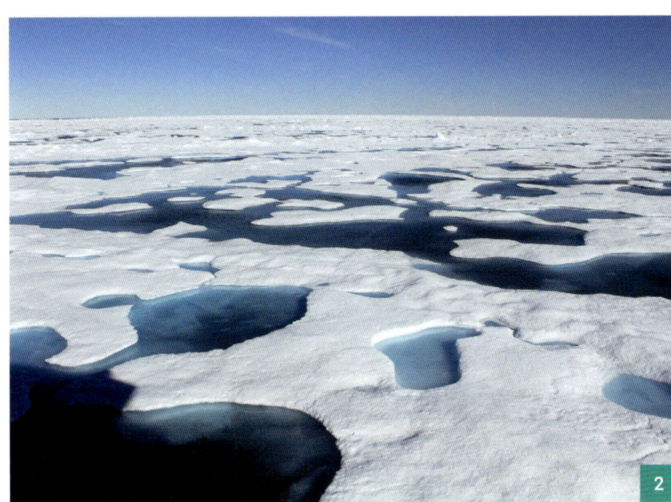

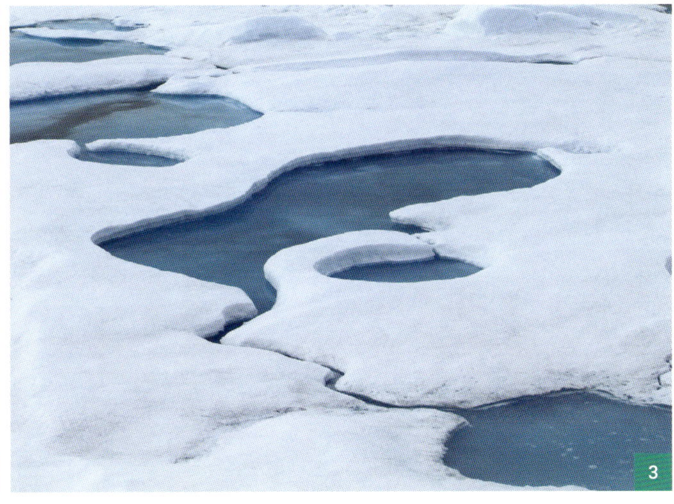

A melting pond and the ARAON in the Arctic (photo. by C.W. Rim) ▾

4

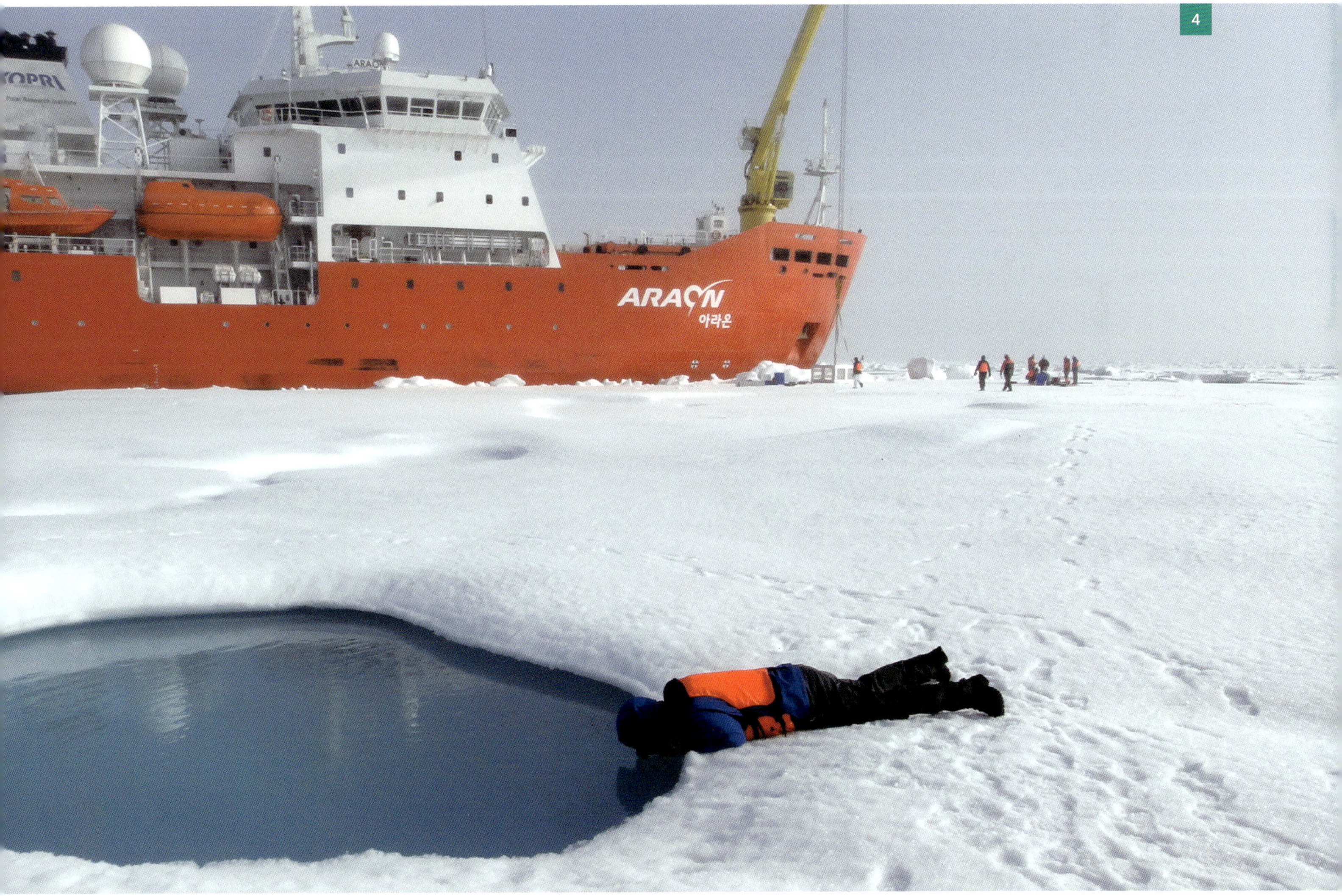

Aerial View of Pack Ice

The icebreaking research vessel ARAON uses helicopter to find a route through which the vessel must pass or to find an ice sheet of a suitable size for an icebreaking test.

1 3 Beaufort Sea

2 4 Amundsen Sea

▼ IBRV ARAON in the Beaufort Sea

1

2

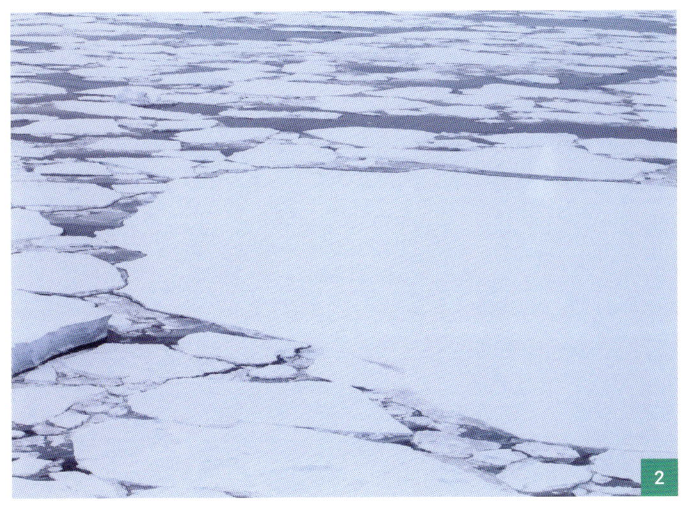

Big ice floe selected for icebreaking test ▲

3

Pack ice ▲

Path of ARAON ▼

4

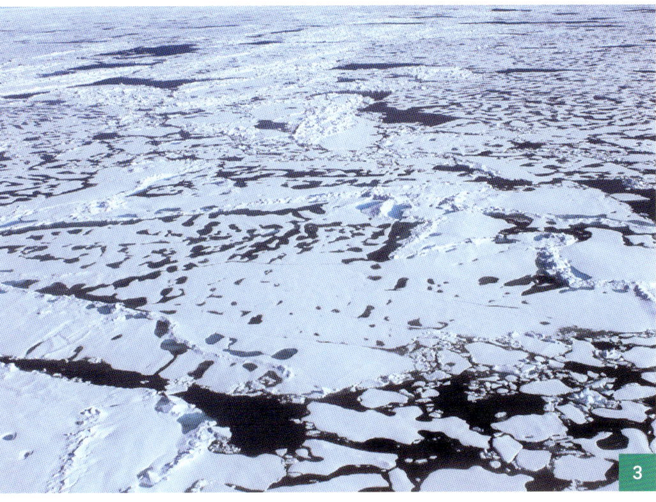

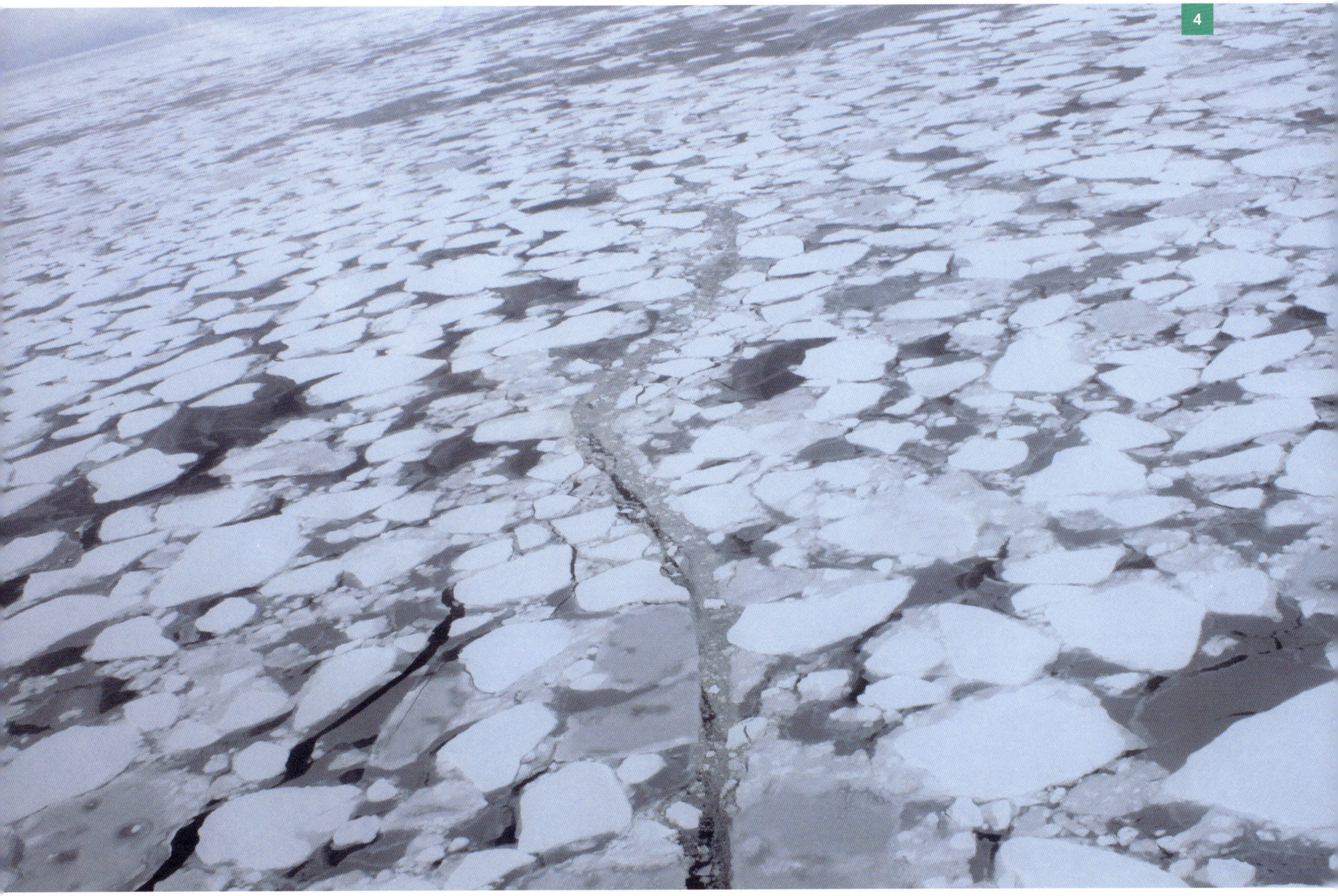

Aerial View of Pack Ice

1 ~ 4 Amundsen Sea

▼ ARAON in pack ice

1

ARAON in pack ice ▲

Aerial view of Antarctic pack ice ▲

Aerial view of Antarctic pack ice and a tabular iceberg ▼

Aerial View of Pack Ice

1 ~ 4 Amundsen Sea

▼ Aerial view of Antarctic pack ice

1

Aerial view of Antarctic pack ice and icebergs ▼

4

Unusual Colors and Shapes of Sea Ice

Floating sea ice with unusual shapes and colors is often found. Usually, algae attach to and grow on the bottom of old ice that is more than 2 years old, so the color of the overturned ice sheet is green or brown in some cases. Occasionally, soil or dust on land is moved by wind and deposited on ice sheets.

1 ~ 4 Beaufort Sea

▾ Bluish color of an old ice among the dirt-covered ice

1

Sea ice and its subsea part ▲

Sea ice and its subsea part upside down ▲

Sea ice and its subsea part ▼

Unusual Colors and Shapes of Sea Ice

1 ~ 3 Amundsen Sea

4 Beaufort Sea

▾ Unusual shape of old sea ice

1

A sea ice piece in wave ▲

Unusual shape of old sea ice ▲

Brown color of a broken ice piece and accumulation of algae ▼

Unusual Colors and Shapes of Sea Ice

1 ~ 4 Beaufort Sea

▾ Brown and green color of broken ice pieces

1

Brown and green color of broken ice pieces ▲

Dirt-covered ice ▲

Brown and green color of broken ice pieces ▼

Part 3.

Land-based Ice and Icebergs

Glacial Ice on Land

Glaciers

Ice found in polar regions can be divided into ice formed from glaciers on land and ice formed by freezing sea water, depending on the source of the formation. A glacier is formed from snow that has accumulated on land for hundreds of thousands of years and is pressed under its own weight into solid ice. At a depth of 10 m close to the surface, the specific gravity of firn, the stage in which snow changes to ice, is about 0.6, but at a depth of 200 m it becomes 0.91, which is close to pure ice. A glacier that is large enough to cause a fluid-like viscous flow with its own weight alone due to the accumulation of snow for a long time slowly moves to a lower terrain according to gravity. When this glacier reaches the shore, the ice at the tip of the glacier falls into the sea, which is an iceberg.

1 4 Glennallen, Alaska
2 3 Anchorage, Alaska

▼ Inland glacier near Glennallen, Alaska

1

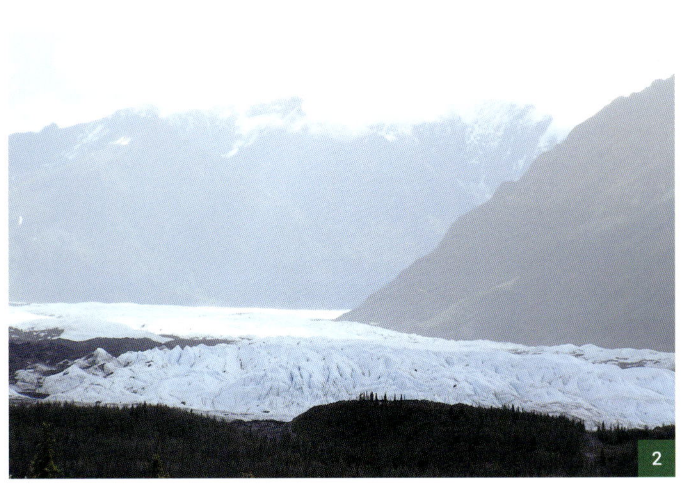

2

Matanuska glacier ▲

3

Matanuska glacier ▲

Inland glacier ▼

4

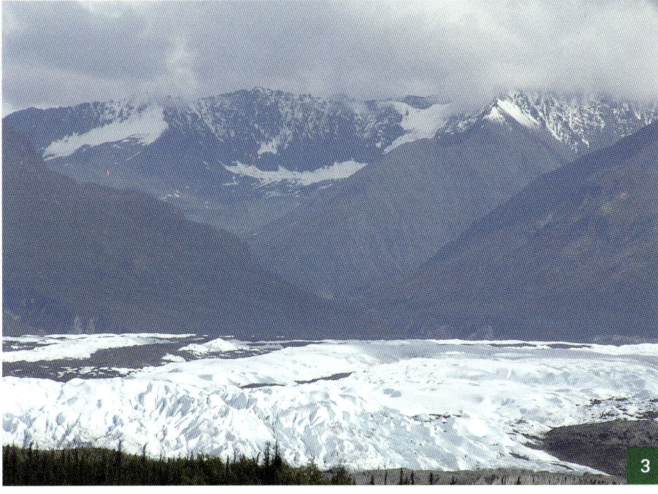

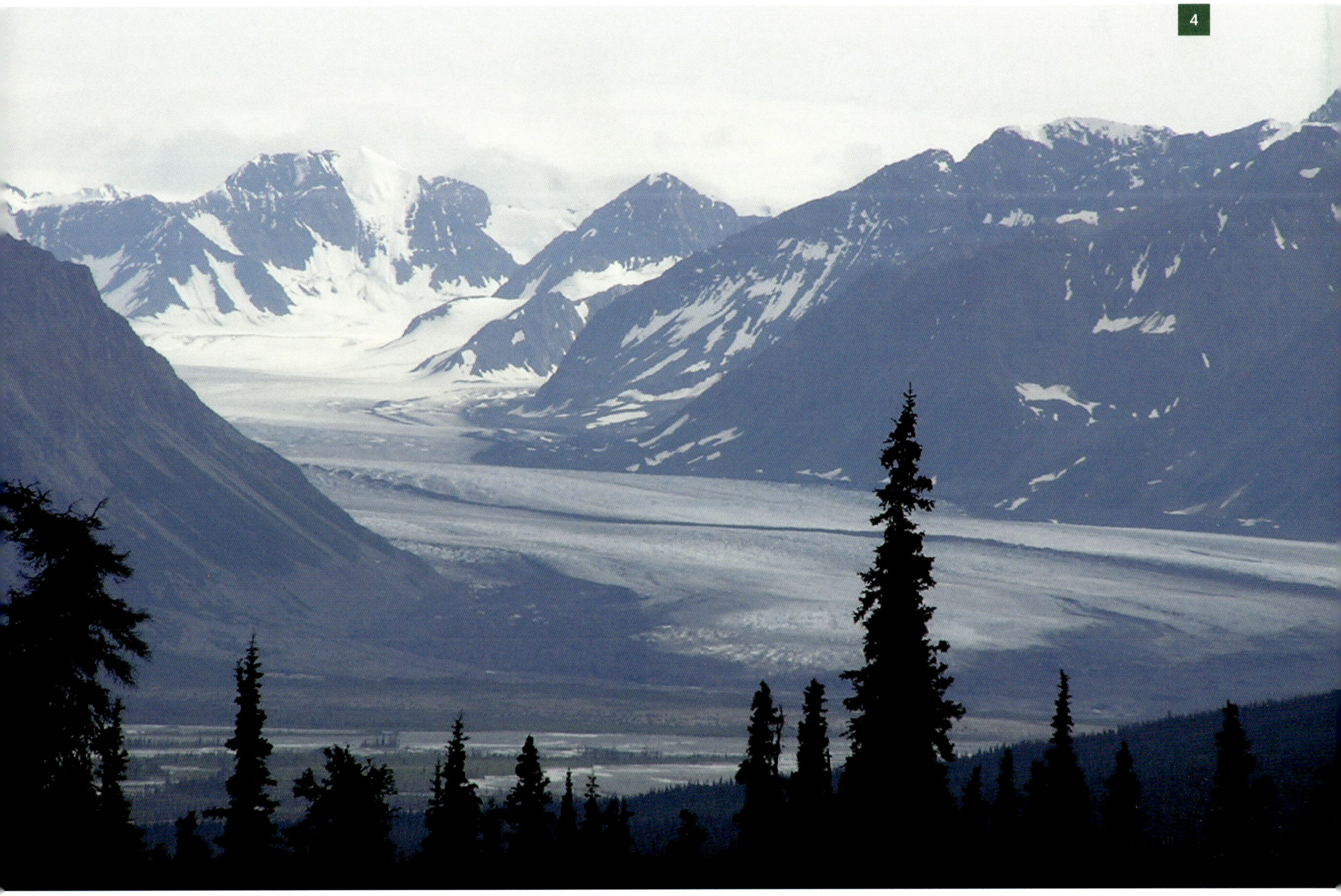

Glacial Ice on Land

Glaciers

1 4 Whittier, Alaska

2 3 Anchorage, Alaska

▼ Coastal glacier near Whittier, Alaska

1

2

Matanuska glacier ▲

3

Matanuska glacier ▲

Coastal glacier ▼

4

Glacial Ice on Land

Snow and Ice on Land's Ends

1 4 Amundsen Sea

2 3 Whittier, Alaska

1

A glacier's end making small bergy bits ▲

Antarctic glacial ice in the Amundsen Sea ▼

Glacial Ice on Land

Snow and Ice on Land's Ends

1 ~ 4 Amundsen Sea

▾ Snow and ice on Antarctic land

1

2

3

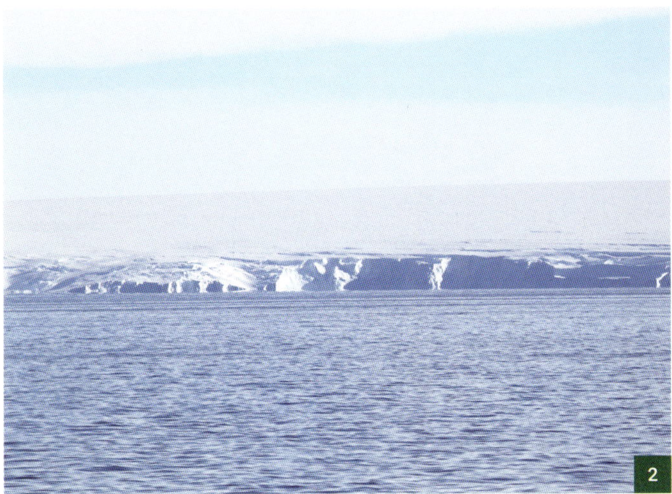

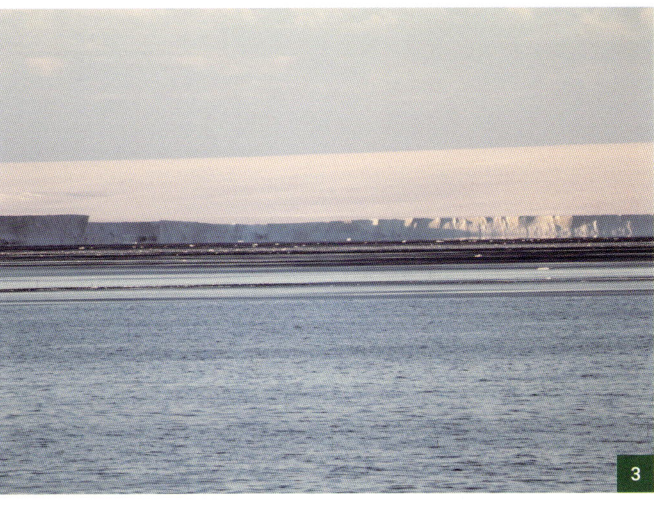

Snow and ice on Antarctic land ▼

4

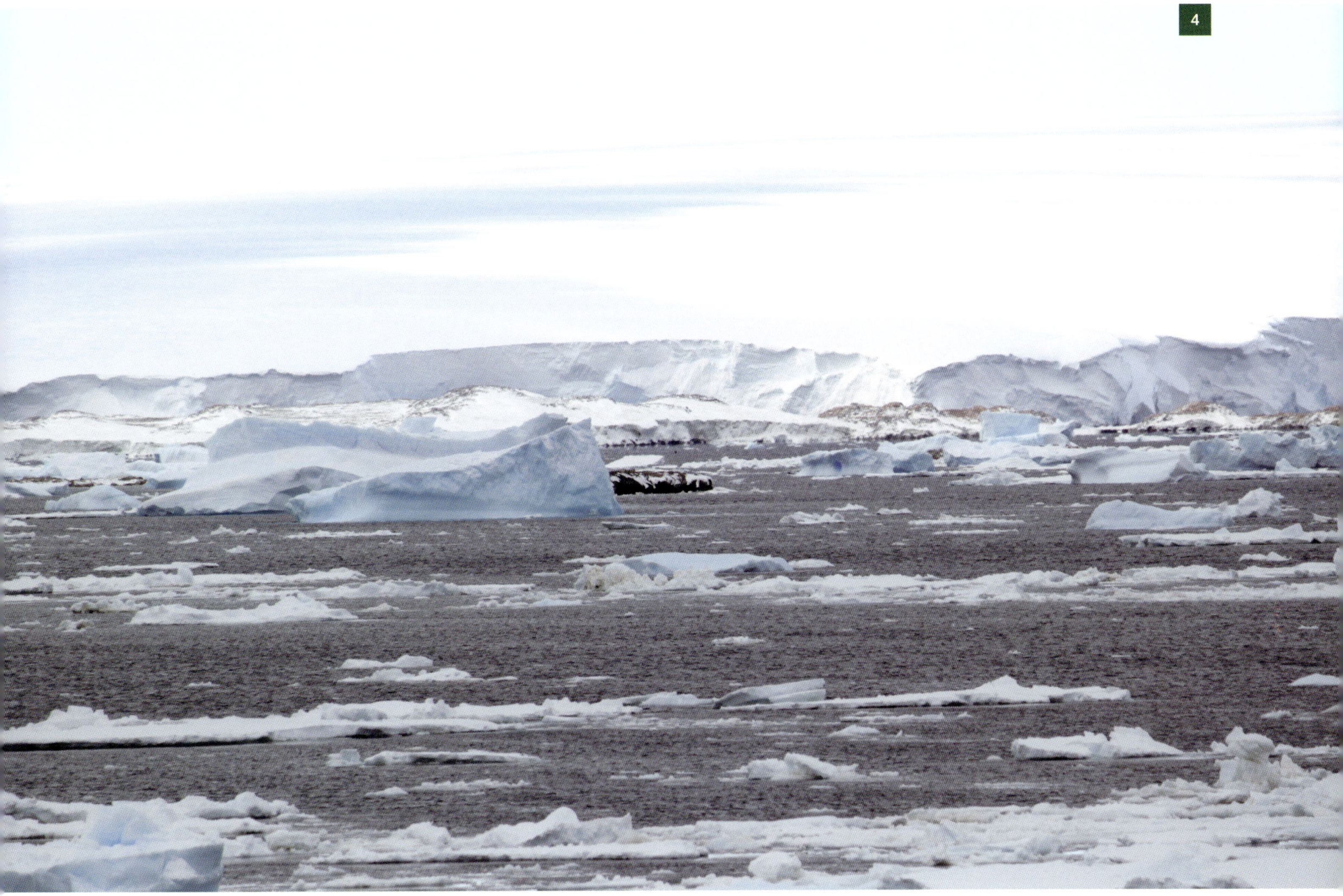

Glacial Ice on Land
Snow and Ice on Land's Ends

1 ~ 4 Amundsen Sea

▾ Antarctic glacial ice

2

Antarctic glacial ice ▲

3

Snow and ice on Antarctic land ▲

Antarctic glacial ice ▼

4

Ice Shelves

In Antarctica, which is covered with glaciers with an average thickness of 2,000 m or more, the tip of the glacier is connected to the sea. It is usually spread widely in the horizontal direction, and the surface is flat or has gentle undulations. Each year, land-based glaciers move to the sea and grow accordingly, and in some places where the water is low along the coast, the bottom surface is attached to the sea floor.

1 Pine Island Bay, Amundsen Sea

2 ~ 4 Amundsen Sea

▼ Ice shelves in the Pine Island Bay, Amundsen Sea

4

Ice Shelves

1 ~ 4 Amundsen Sea

1

2

3

A drain hole in the wall of ice shelves ▼

4

Ice Shelves

1 ~ 4 Amundsen Sea

1

2

3

4

Icebergs

An iceberg is a mass of ice that has fallen into the sea from a glacier on land, and its source is fresh water ice. The specific gravity of the iceberg is 0.85 to 0.91, so what we see at sea level is only a fraction of the iceberg, and most of the iceberg is hidden beneath the surface. Icebergs are formed mainly on the western coast of Greenland in the northern hemisphere and on the ice cap of Antarctica in the southern hemisphere. Icebergs are divided into two types according to their shapes.

Blocky Iceberg This iceberg is separated from the glacier on land and fell directly into the sea. It is usually a rectangular parallelepiped with a thickness to width ratio of 1:1 to 1:2 and is about 50-500 m in size.

Tabular Iceberg In Antarctica, when glaciers move from land to sea, they often form vast areas of land-connected ice sheets. Usually the ratio of thickness to width is 1:10 or more, but sometimes super-large icebergs with a length of 160 km and a thickness of 500 m are found.

Icebergs can survive for years to completely melt, and while they are drifting, they are transformed into a variety of shapes by warm sea water and temperatures.

Dome Iceberg The submerged portion of the iceberg melts so badly that the iceberg becomes unstable, and when it is turned over, a gently molten round surface is revealed.

Pinnacle Iceberg When the upper part of the water surface is severely eroded by sunlight and wind, it becomes a pyramid-shaped or very irregular spire-shaped iceberg.

1

1 Amundsen Sea

2 ~ 3 Southern Ocean

2

3

A blocky iceberg, Amundsen Sea ▼

Icebergs

Blocky Icebergs

1 Southern Ocean

2 ~ 4 Amundsen Sea

1

2

3

4

Icebergs

Blocky Icebergs

1 Amundsen Sea

2 ~ 4 Southern Ocean

1

2

3

A blocky iceberg in the Southern Ocean ▼

4

Icebergs

Blocky Icebergs

1 ~ 3 Amundsen Sea

4 Ross Sea

1

2

3

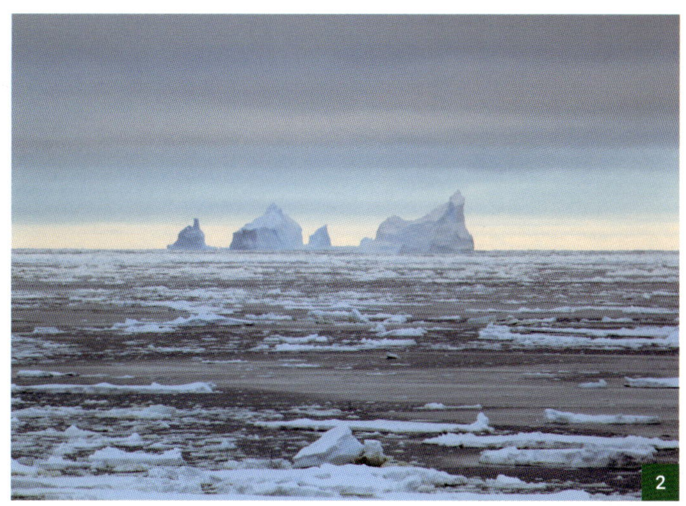

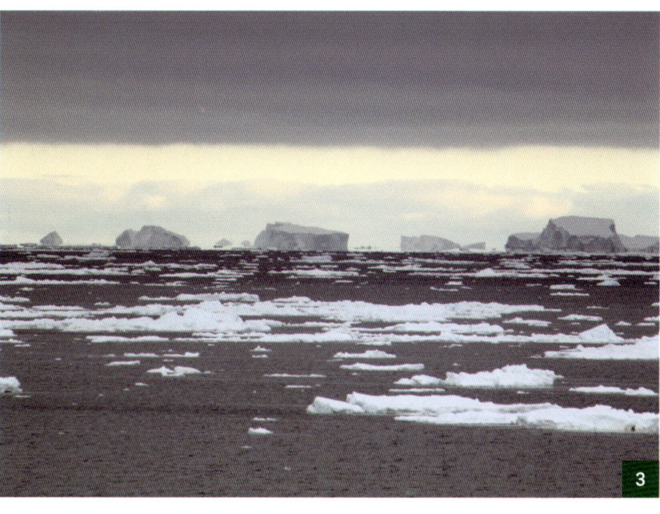

A blocky iceberg, Ross Sea ▼

4

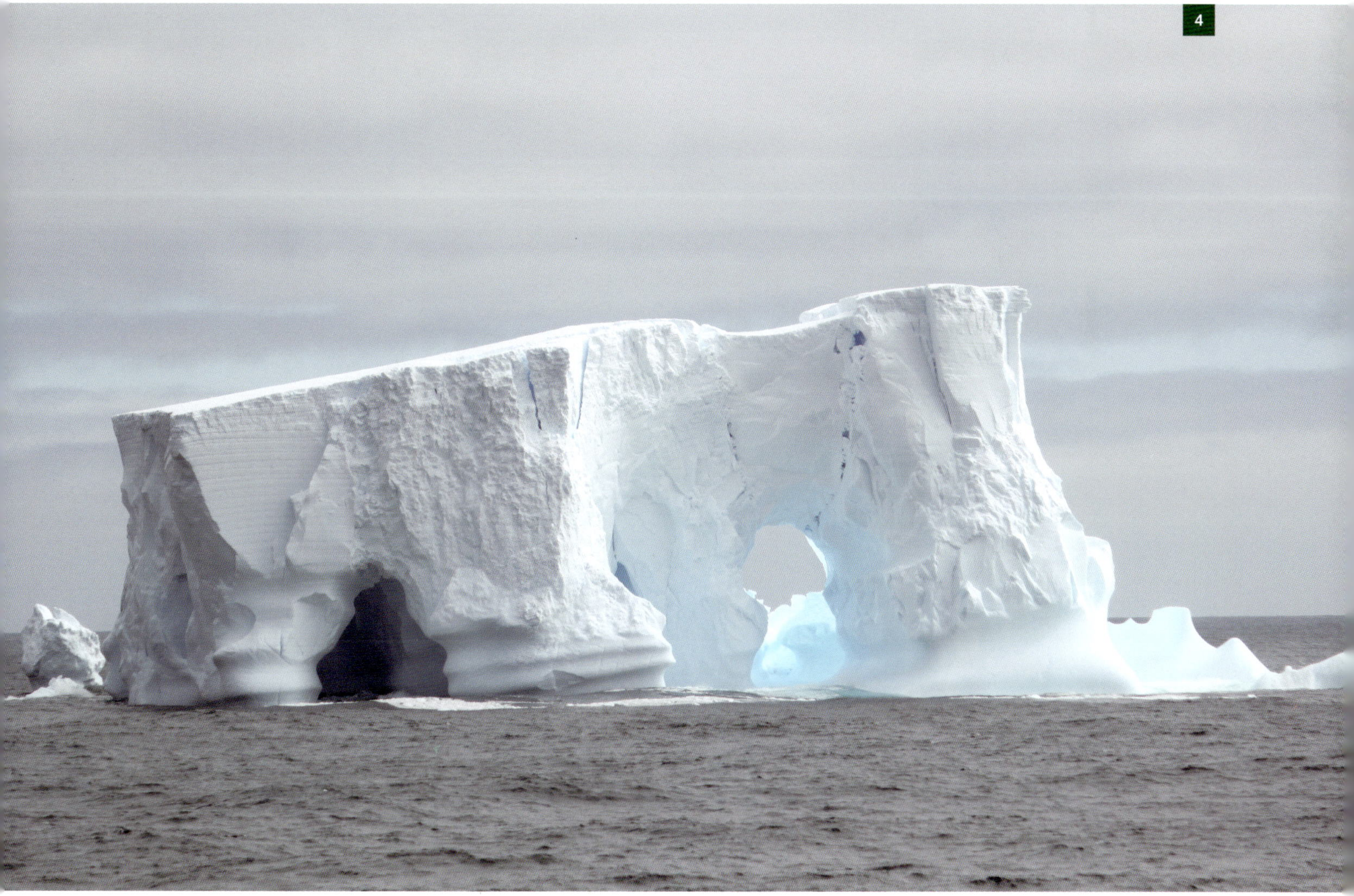

Icebergs

Tabular Icebergs

1 ~ 4 Amundsen Sea

▼ A tabular iceberg found in the Amundsen Sea

1

4

Icebergs

Tabular Icebergs

1 ~ 4 Amundsen Sea

1

2

3

A tabular iceberg and a half moon, Amundsen Sea ▼

4

Icebergs

Tabular Icebergs

1 ~ 4 Amundsen Sea

1

4

Unusual Features

Icebergs that have fallen from the upper layer of the glacier have a small specific gravity and contain a lot of snow crystals and air bubbles inside, so they appear white from a distance. However, since the ice formed below the inner glacier is compacted with high pressure, the air bubbles are compressed into smaller sizes and appear transparent. From a distance, these icebergs have a very beautiful blue color. However, among the various types of icebergs floating in the sea, some with unusual shapes and colors are sometimes found. A typical case is when volcanic ash was deposited on the glacier layer during a volcanic eruption when glaciers were formed on land long ago, and a black deposit layer was formed.

1 2 Cape Burks, Amundsen Sea

3 4 Amundsen Sea

▾ An iceberg with black deposit layers

1

2

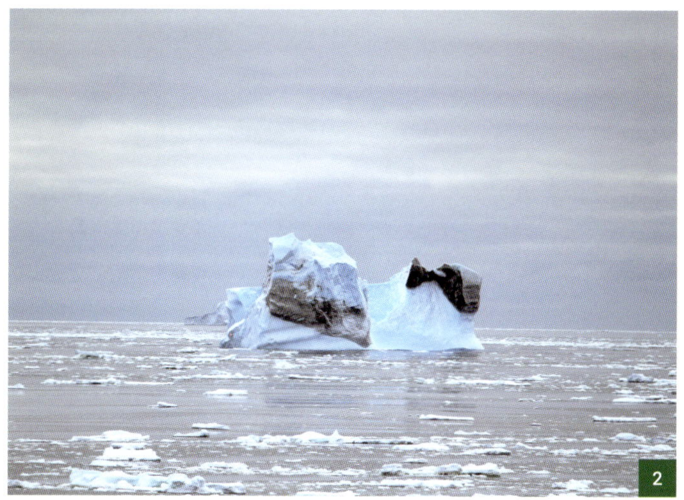

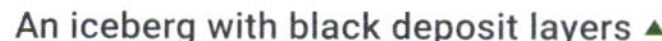

An iceberg with black deposit layers ▲

3

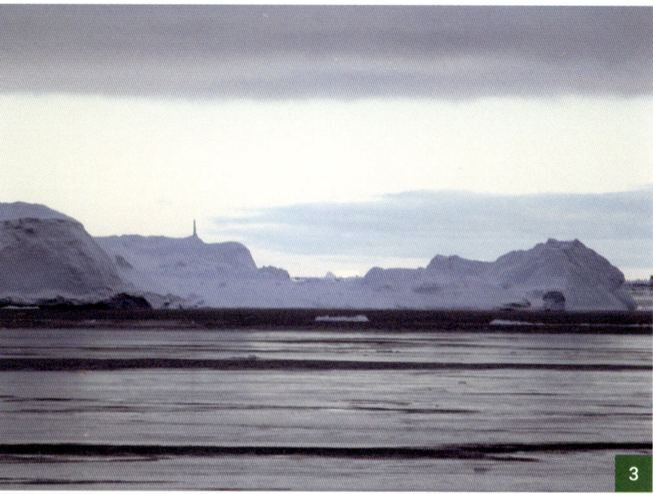

A needle-shape found on top of an iceberg ▲

Unusual combination of blue and black colors ▼

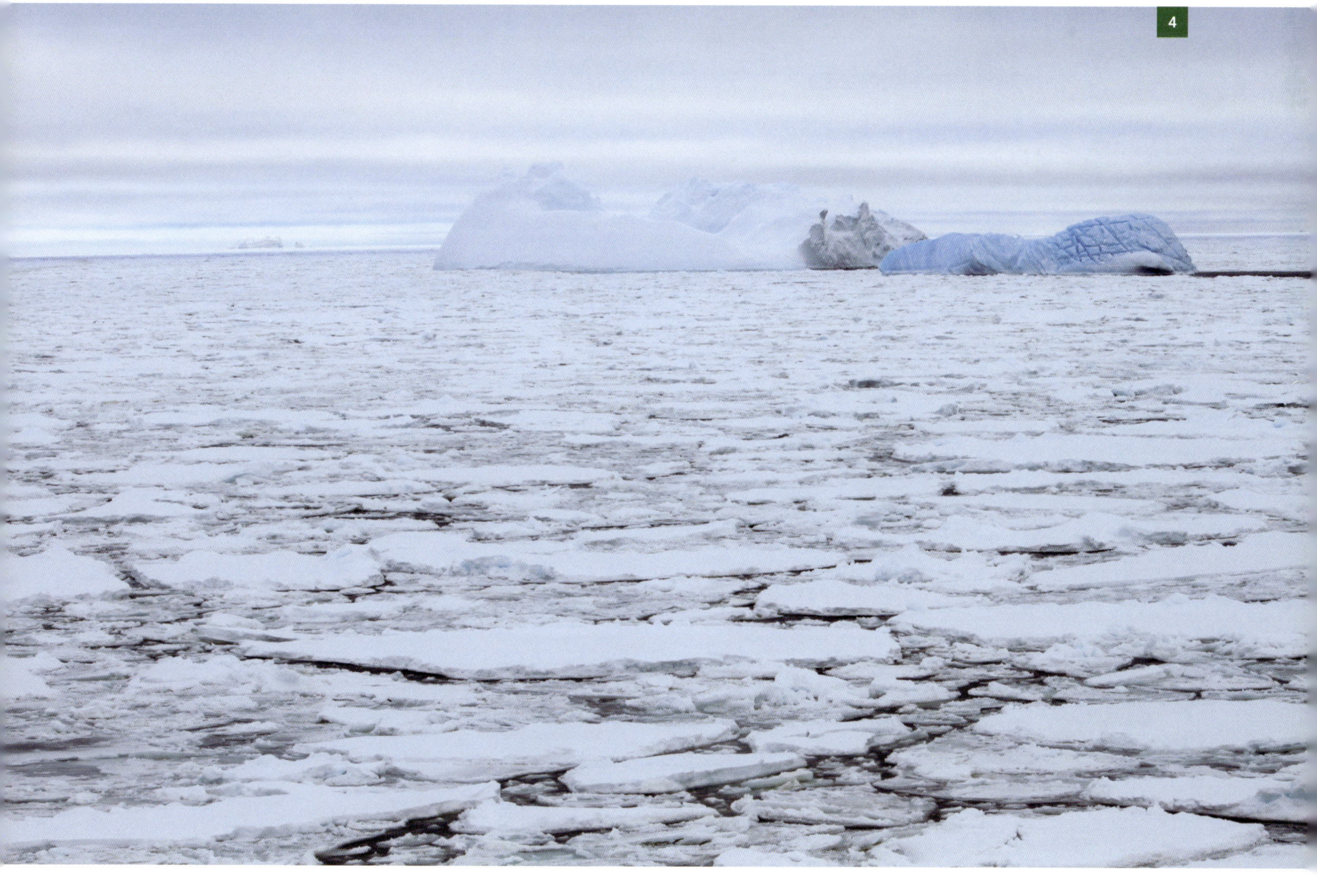
4

Unusual Features

1 ~ 4 Amundsen Sea

▾ Unusual features of a dome type iceberg

1

2

3

4

Unusual Features

1 ~ 4 Amundsen Sea

1

2

3

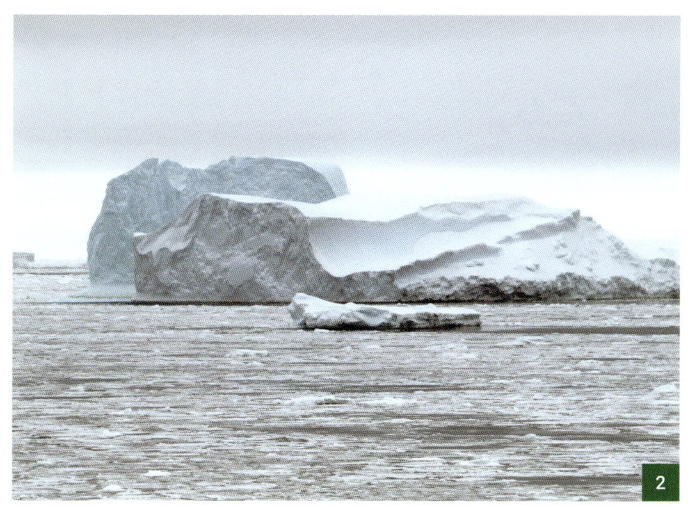

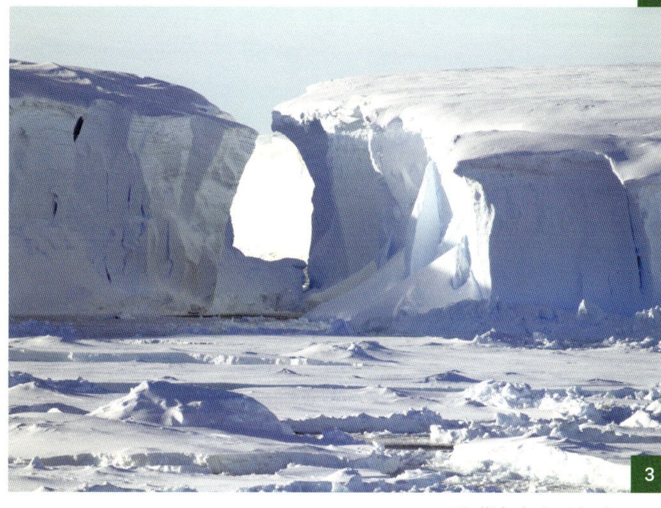

Collided giant icebergs ▲

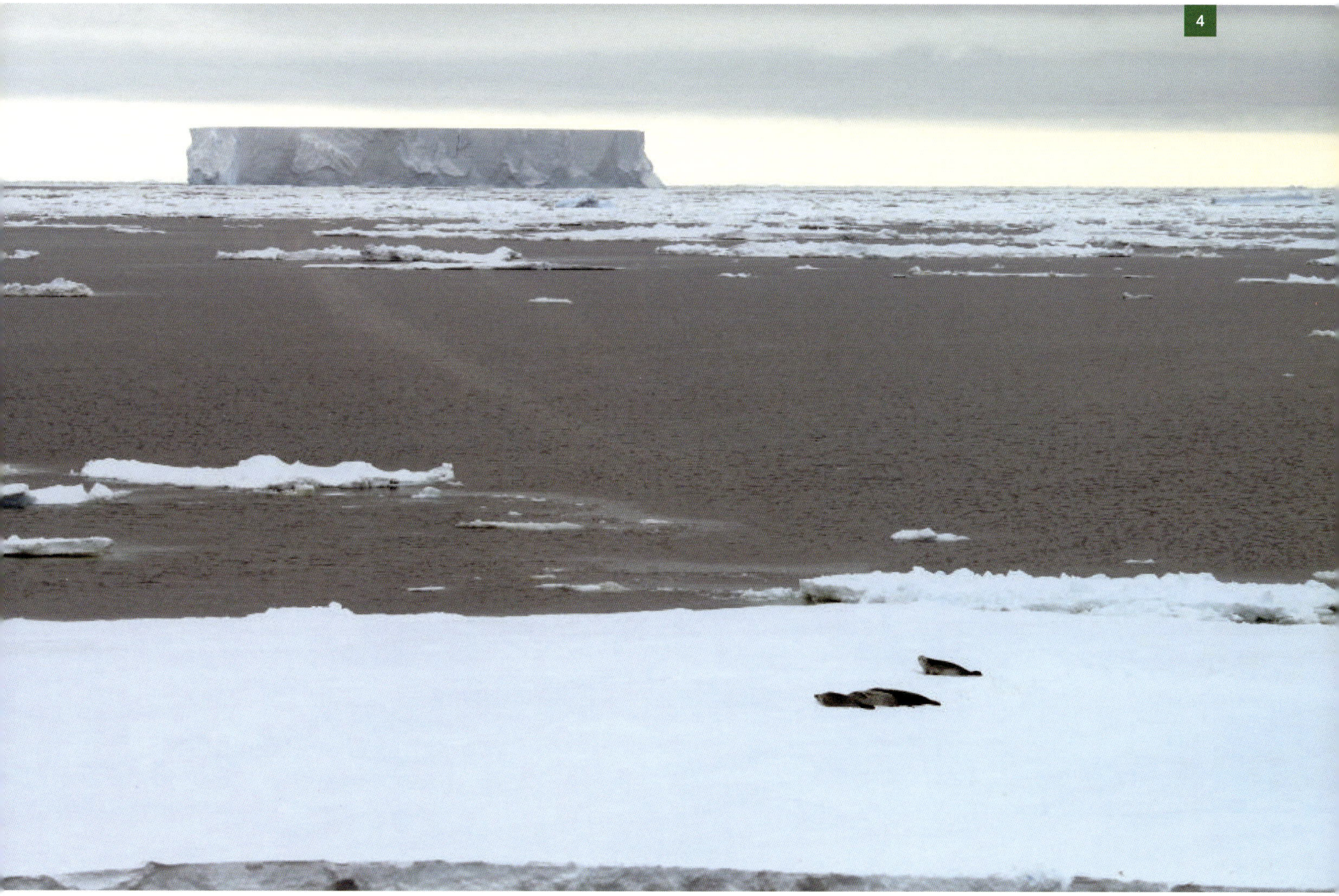

4

Part 4.

Icebreaking Patterns

Cracks in Ice

Polar ice is a multi-year ice in which freezing and thawing are balanced, and is not an integrated ice mass, but an aggregate of ice sheets split by open leads. In the Arctic Ocean around the time when ice melts and disappears in summer, the Arctic sea ice continues to move due to the flow of major ocean currents such as the Beaufort Gyre. Numerous drift ice forms a floating channel. In the case of the Amundsen Sea in Antarctica, except for the existence of a huge iceberg, it is in a state of drift ice similar to that of the Arctic Ocean. Although fragments of multi-year ice appear to be frozen and united, it is difficult to find one that actually exists as a large, unbroken ice sheet. It can be seen that the movement of drift ice is much more dynamic. On the other hand, the movement of the icebreaker causes cracks in the ice sheet. The occurrence of fast-propagating cracks serves to reduce the energy required to break the ice.

1 Amundsen Sea
2 ~ 4 Beaufort Sea
3 Baltic Sea

1 A major crack generated by the action of the IBRV ARAON
2 4 A major crack among puddles
3 A crack generated by the action of icebreaker SAMPO

1

2

3

A major crack among puddles ▼

4

Cracks in Ice

1 Beaufort Sea

2 ~ 4 Amundsen Sea

▾ A major crack among puddles

1

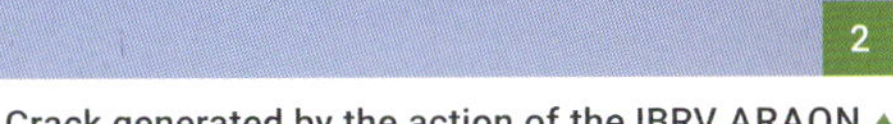
Crack generated by the action of the IBRV ARAON ▲

An open lead generated in the newly frozen ice ▲

Crack generated by the action of the IBRV ARAON ▼

Broken Pieces and Cusps

In thin ice, radial cracks are mainly formed at the bow by the movement of the icebreaker, but in the case of thick ice, a cusp is created in which the ice fragments are broken in a circle like a half moon on the side of the bow.

1 4 Beaufort Sea
2 3 Baltic Sea

▾ Cusps seen from the side of the IBRV ARAON

1

2

3

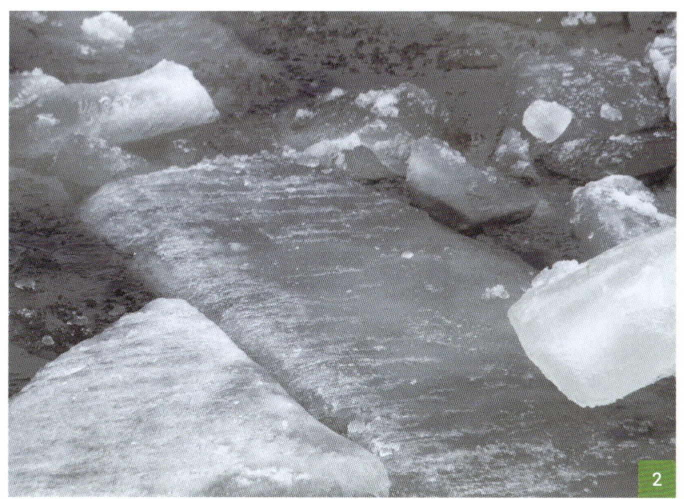

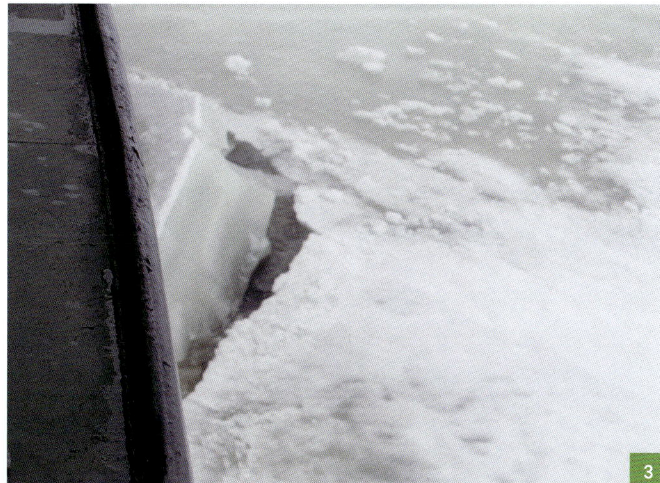

Broken sea ice pieces by icebreaker SAMPO ▲

Relatively thin ice sheet and cusps ▼

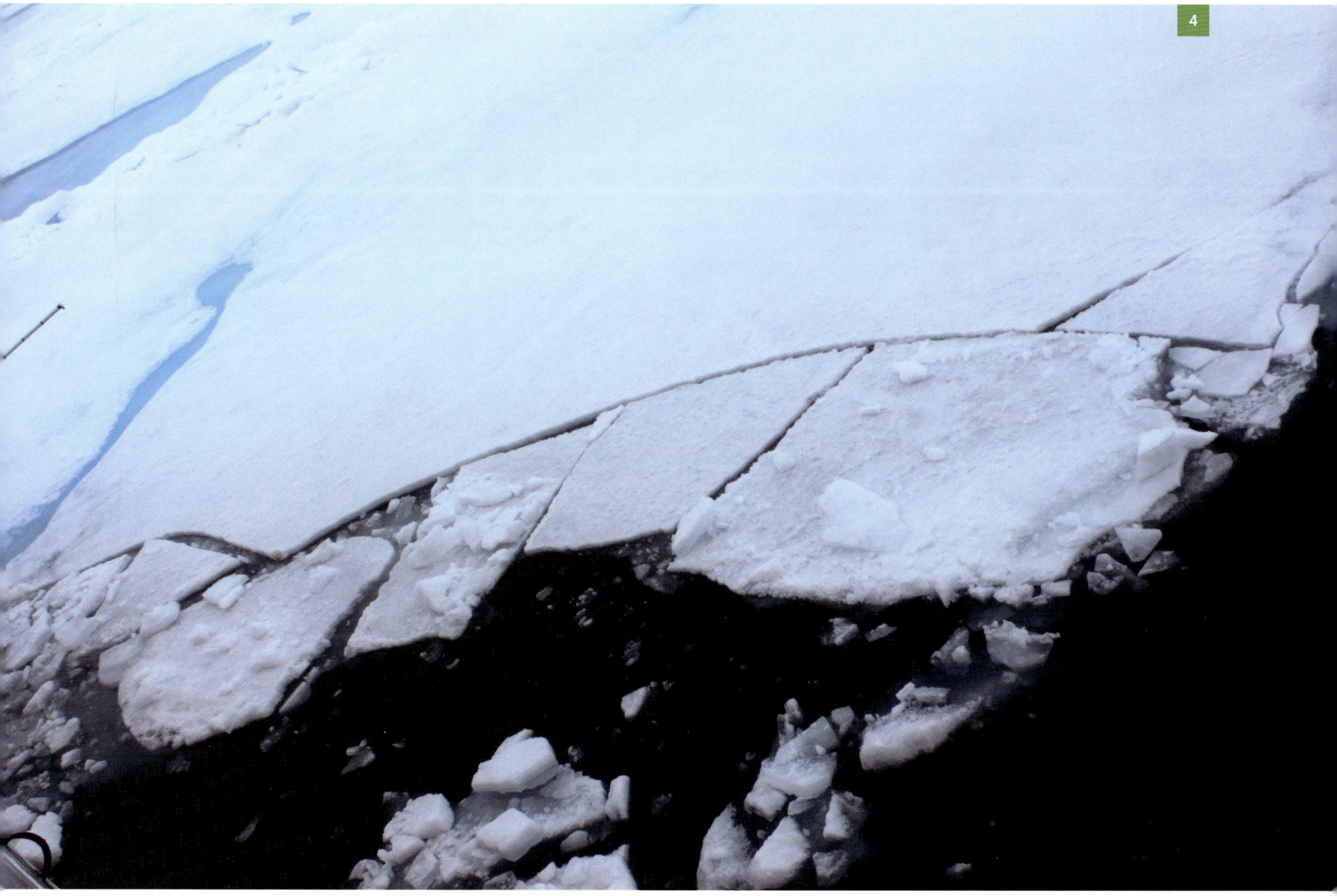
4

Broken Pieces and Cusps

1 4 ~ 6 Amundsen Sea

2 3 Beaufort Sea

▾ Broken ice pieces during icebreaking test

1

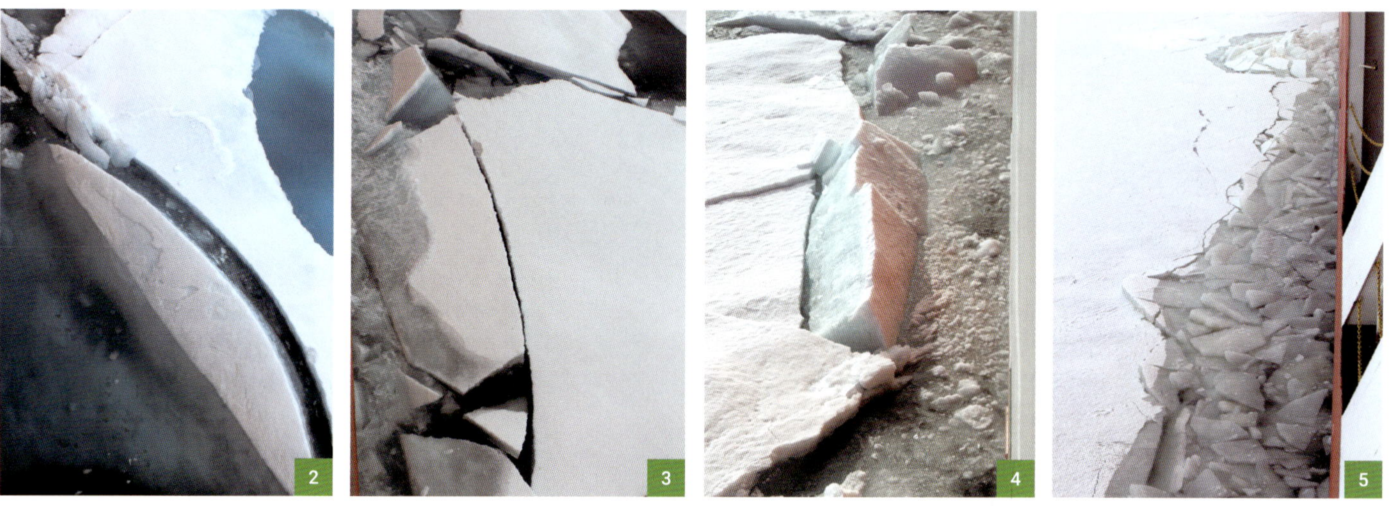

2 3 Cusps seen from the side of the IBRV ARAON

4 5 Broken ice pieces during icebreaking test

Broken ice pieces during icebreaking test ▼

6

Ice Channels

When icebreakers have passed through after breaking flat ice or pack ice, it is very rare that the ice fragments are completely removed, and the waterways are often refilled with broken ice pieces over time. In extreme cold conditions, the waterway created by the preceding icebreaker quickly re-freezes and the vessel following is often trapped in the ice.

1 4 Amundsen Sea

2 3 Baltic Sea

▾ Open ice channel in the Antarctic pack ice behind the IBRV ARAON

1

Re-closed ice channel ▲

Ice channel made by the Finnish icebreaker SAMPO ▲

Open ice channel in the Antarctic pack ice behind the IBRV ARAON ▼

Ice Channels

1 ~ 4 Amundsen Sea

▾ Open ice channel in the Antarctic pack ice behind the IBRV ARAON

1

2 3 4 Open ice channel in the Antarctic pack ice behind the IBRV ARAON

Buoy deployment work along the open lead made by ARAON in the Amundsen Sea ▼

4

Part 5.
Experimental Activities on Sea Ice

Coring and Measuring Ice Properties

Research on sea ice involves an icebreaking test that directly breaks ice using an icebreaker to estimate the magnitude of ice load to determine how much force a ship will receive when breaking the ice. In general, material properties of sea ice are also measured during icebreaking test. It is the task of measuring the temperature, salinity, density and compressive strength according to depth by drilling a hole and collecting ice specimens.

▾ Sea ice research team working on Arctic sea ice

1

Extracted ice core samples from a sea ice floe ▲

Compressive strength test with an ice specimen ▲

Cutting sea ice core to make a test specimen ▼

Coring and Measuring Ice Properties

▾ Electromagnetic device to measure ice thickness

1

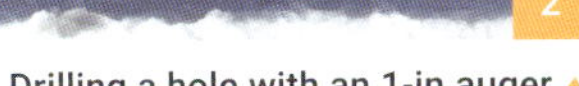

Drilling a hole with an 1-in auger ▲

Drilling sea ice with 3-in corer ▲

Participated members in 2012 Amundsen Sea Expedition ▼

Icebreaking Research Vessel ARAON

The IBRV ARAON was built to support polar research and transport for the King Sejong Station and Jangbogo Station in Antartica. Since its construction in 2009, the ARAON has been conducting research on polar seas and ice by visiting the Antarctic and Arctic Oceans once a year. The 109 m long and 6,930 gross tonnage ARAON is an icebreaker Class 10 of the Korean Register of Shipping POLAR class and has the ability to break a 1 m thick ice sheet at a speed of 3 knots. A maximum of 65 researchers can be on board.

▾ Jangbogo Station, Antarctica

1

2

Airial view of the IBRV ARAON shown behind a huge iceberg ▲

3

Front view of the IBRV ARAON ▲

IBRV ARAON stopped for an icebreaking test at 1m level ice sheet, Amundsen Sea ▼

4

Activities Using Helicopters and Snowmobiles

In the polar regions, helicopters or snowmobiles (ski-doo) are used to transport goods over long distances or areas that are difficult to move by ships. ARAON can carry two small helicopters and this helicopter are used to conduct reconnaissance activities to open the route of the ARAON surrounded by ice or to find a suitable size ice sheet for sea ice research. Snowmobiles are efficient when moving or moving goods on a relatively flat and large area of ice.

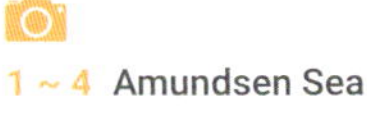
1 ~ 4 Amundsen Sea

▾ A helicopter transports scientists and equipments

1

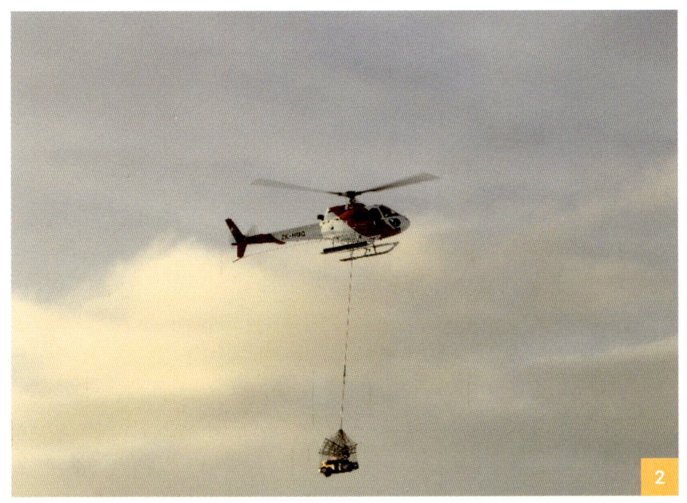

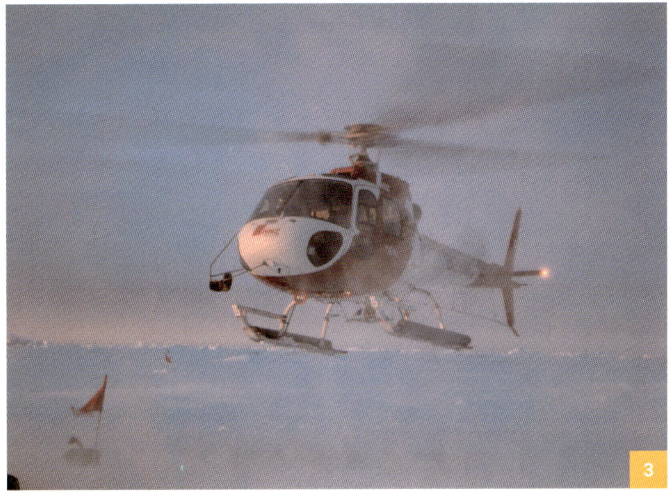

2

3

Transportation of test equipments ▲

Transportation of test equipments ▼

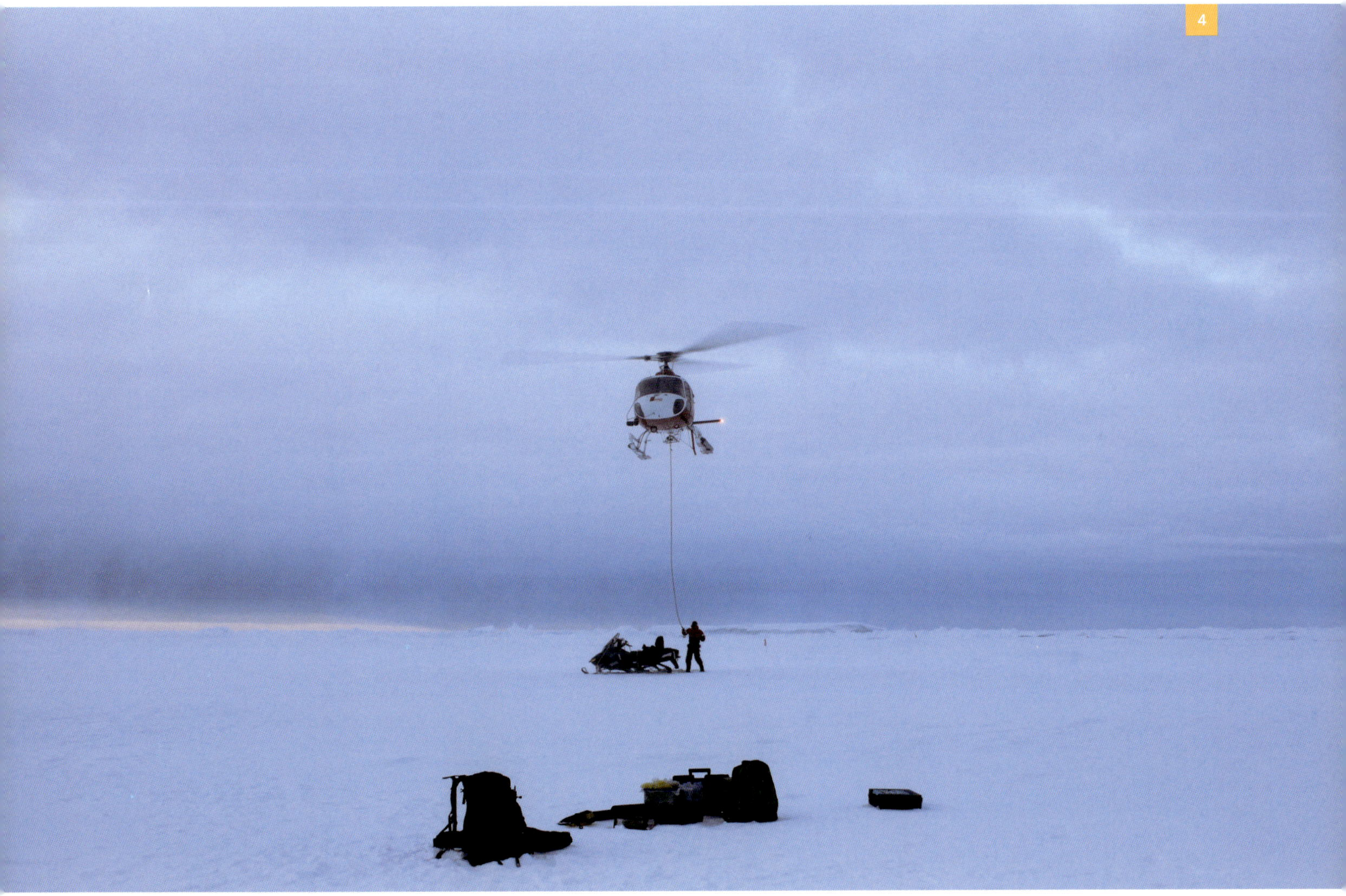

4

Activities Using Helicopters and Snowmobiles

▼ Transportation by snowmobiles, Svalbard

1

2 3 4 Transportation by snowmobiles

Part 6.

Animals on Sea Ice

The representative animals of the Arctic are polar bears and seals, but in addition, reindeer and bighorn deer are migrating, and indigenous peoples including Inuit also live along the coast. Mosses live along the coast in Antarctica, and animals such as penguins, seagulls, and Antarctic terns live in Antarctica. Krill, sea lions, and whales inhabit the Antarctic Ocean. The cold and dry Antarctic highland is an environment where no living things can survive and no trees or grass can grow.

Svalbard Reindeer

Svalbard reindeer are ubiquitous like the city's feral cats. They don't care as much about the camera's approach as they often had contact with people. They mainly eat grass that grows in a short period of summer. Svalbard's reindeer evolved according to the climate of the Arctic Circle.

▾ Svalbard reindeer (male) resting near Longyearbyen

1

2 3 Svalbard reindeer (female) found near Longyearbyen

Polar Bears

Polar bears are the top predators of animals in the Arctic Circle. Polar bears' favorite food is swarms of sea lions, but as the ice that polar bears can stay on is gradually disappearing due to global warming, polar bears have no choice but to swim far away and move to places where there is ice. In the 2010 Arctic Sea Expedition, polar bears were observed hundreds of kilometers from land. The polar bear was frightened by the sound of the helicopter and showed a back view.

Polar bears are very dangerous creatures that even attack humans, especially when they travel far into the sea due to lack of food. Polar bears have a good sense of smell, so they can smell well from a distance and attack by suddenly climbing up on the ice from behind in search of food, so be very careful when working on sea ice. As the ARAON docks next to the ice in the Arctic sea and the researchers step onto the ice, a watchman with a gun will keep a look around the workers from the deck, preparing for a possible polar bear appearance.

▾ A polar bear appeared during Arctic sea ice reconnaissance

1

2

Foot print of a polar bear ▲

3

A polar bear appeared during Arctic sea ice reconnaissance ▲

Foot print of a polar bear ▼

4

Polar Bears

1 ~ 4 Beaufort Sea

▾ A polar bear appeared during Arctic sea ice reconnaissance

1

2

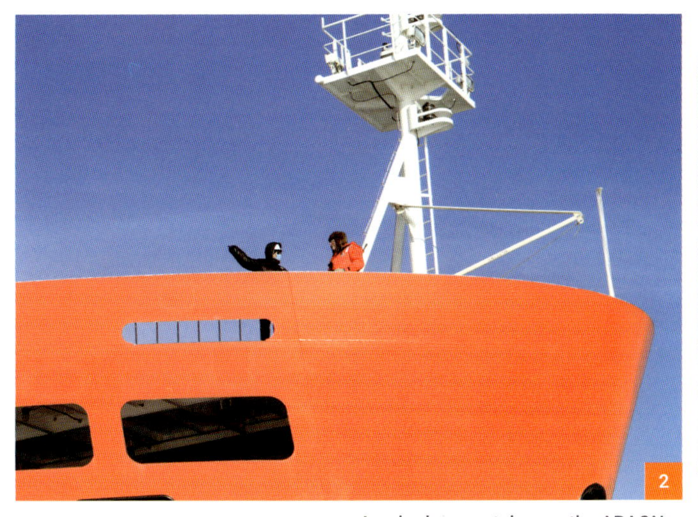

A polar bear watcher on the ARAON ▲

3

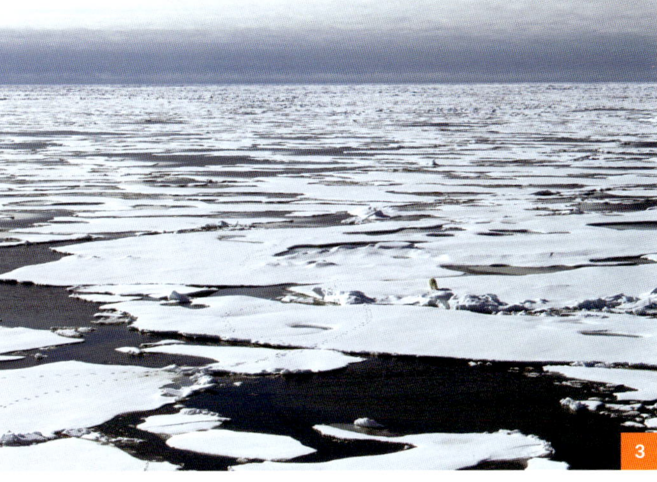

Swimming polar bear ▼

4

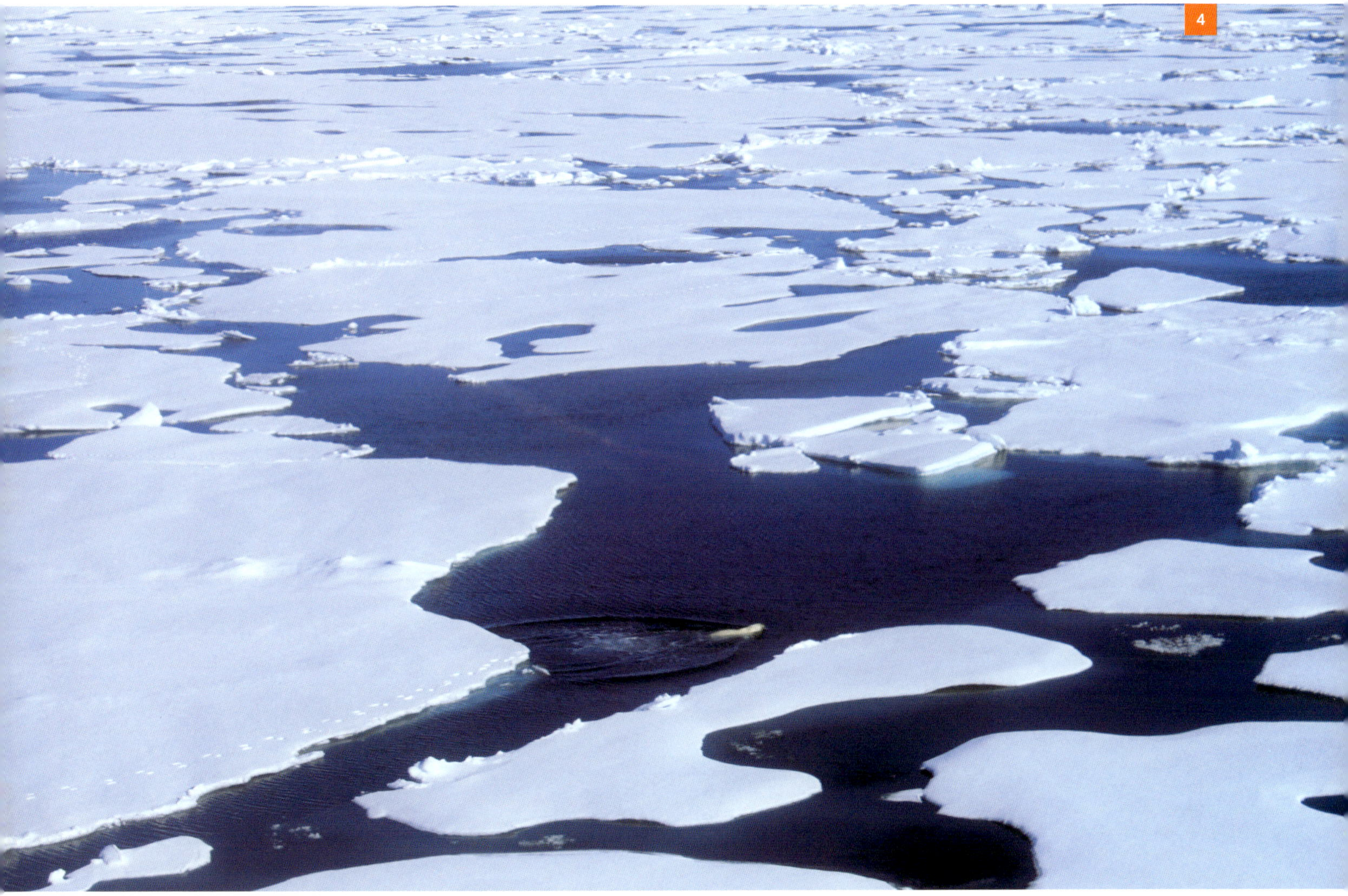

Antarctic Seals

The Antarctic seal which is frequently observed in Antarctica, is a type of marine mammal with flipper-footed feet and can dive for a long time, but must come out of the water to breathe. It climbs on land or on ice to rest, but its anterior fin is small and its anal fin cannot be turned forward, so it crawls to move.

1 2 4 Amundsen Sea
3 Ross Sea

▾ Antarctic seals found in the Amundsen Sea

1

An Antarctic seal in the Ross Sea ▲

An Antarctic seal in the Amundsen Sea ▼

Penguins

The flightless bird, penguin, is known as the representative bird of Antarctica because it inhabits the southern hemisphere centered on Antarctica. The penguin's feathers are highly waterproof, allowing air to be trapped between the feathers and the skin, and a thick layer of fat is developed between the skin and the muscles to block the cold and store energy. Penguins are dull on land, but very agile and graceful in the water have swimming ability. Webbed feet, a triangular tail that acts as a rudder, and powerful muscle-powered wings in harmony, it can move through the water with tremendous propulsion.

The penguins observed during sea ice research in the Amundsen Sea are mainly emperor penguins and Adelie penguins. In addition, gentoo penguins and chinstrap penguins inhabit the coast of Antarctica and islands around the Antarctic Peninsula. Emperor penguin is 110 cm tall and weighs about 30 kg. It can be distinguished by its triangular, light golden hair that runs from the neck to the head. Adelie penguin is a small species of less than 80 cm tall, with a black face and a gray or light brown beak.

Penguins are very curious, so they are not afraid of human approach. In particular, there are many cases where researchers working on sea ice in white, black and red uniforms judged that they were their colleagues and approached them.

▼ A group of penguins

1

1 ~ 3 Amundsen Sea

2 ~ 3 Emperor penguins

Penguins

1 ~ 4 Amundsen Sea

▼ Emperor penguins approaching to the sea ice research team

1

2

A group of penguins ▲

3

Adelie penguins ▲

A group of penguins ▼

4

Terminology

The World Meteorological Organization (WMO) has compiled and published terms related to sea ice used academically.

Anchor ice : Submerged ice attached or anchored to the bottom, irrespective of the nature of its formation.

Bare ice : Ice without snow cover.

Bergy bit : A piece of glacier ice, generally showing 1 to less than 5 m above sea level, with a length of 5 to less than 15 m. They normally have an area of 100-300 m^2

Beset : Situation in which a vessel is surrounded by ice and unable to move.

Blocky iceberg : A flat-topped iceberg with steep vertical sides.

Brash ice : Accumulation of floating ice made up of fragments not more than 2 m across, the wreckage of other forms of ice.

Compact ice : Floating ice in which the concentration is 10/10 and no water is visible.

Consolidated ice : Floating ice in which the concentration is 10/10 and the floes are frozen together.

Crack : Any fracture of fast ice, consolidated ice or a single floe which may have been followed by separation ranging from a few centimetres to 1 m.

Domed iceberg : An iceberg which is smooth and rounded on top.

Drydocked iceberg : An iceberg which is eroded such that a U-shaped slot is formed near or at water level, with twin columns or pinnacles.

Fast ice : Ice which forms and remains fast along the coast. It may be attached to the shore, to an ice wall, to an ice front, between shoals or grounded icebergs. Vertical fluctuations may be observed during changes of sea level. It may be formed "in-situ" from water or by freezing of floating ice of any age to shore and can extend a few metres or several hundred kilometres from the coast. It may be more than one year old in which case it may be prefixed with the appropriate age category (old, second-year or multi-year). If higher than 2 m above sea level, it is called an ice shelf.

Finger rafting : Type of rafting whereby interlocking thrusts are formed like "fingers" alternately over and under the other. This is commonly found in nilas and in grey ice.

Firn : Old snow which has recrystallized into a dense material. Unlike ordinary snow, particles are to some extent joined together; but, unlike ice, the air spaces in it still connect with each other.

First-year ice : Sea ice of not more than one winter's growth, developing from young ice; 30 cm or greater.

Floe : Any relatively flat piece of ice 20 m or more across. Floes are subdivided according to horizontal extent as follows:

- **Small** : 20-100 m across
- **Medium** : 100-500 m across
- **Big** : 500-2,000 m across
- **Vast** : 2-10 km across

- **Giant** : Greater than 10 km across

Flooded ice : Ice which has been flooded and is heavily loaded by water or water and wet snow.

Frazil ice : Fine spicules or plates of ice suspended in water.

Glacier : A mass of snow and ice continuously moving from higher to lower ground or, if afloat, continuously spreading. The principal forms of glaciers are: inland ice sheets, ice shelves, ice streams, ice caps, and various types of mountain (valley) glaciers.

Glacial ice : Ice in or originating from a glacier, whether on land or floating on the sea as icebergs, bergy bits, growlers or ice islands

Grease ice : A later stage of freezing than frazil ice where the crystals have coagulated to form a soupy layer on the surface. Grease ice reflects little light, giving the water a matte appearance.

Grey ice : Young ice 10-15 cm thick, less elastic than nilas and breaks on swell. It usually rafts under pressure.

Grounded ice : Floating ice which is aground in shoal water.

Growler : Piece of ice smaller than a bergy bit and floating less than 1 m above the sea surface, a growler generally appears white but sometimes transparent or blue-green in colour. Extending less than 1 m above the sea surface and normally occupying an area of about 20 m^2, growlers are difficult to distinguish when surrounded by sea ice or in high sea state.

Hummocked ice : Ice piled haphazardly one piece over another to form an uneven surface. When weathered it has the appearance of smooth hillocks.

Ice cake : Any relatively flat piece of ice less than 20 m across.

Ice concentration : The ratio expressed in tenths describing the area of the water surface covered by ice as a fraction of the whole area.

- **Compact pack ice** : Concentration 10/10, no water visible.
- **Consolidated pack ice** : Concentration 10/10, floes frozen together.
- **Very close pack ice** : Concentration 9/10 to less than 10/10.
- **Close pack ice** : Concentration 7/10 to 8/10, floes mostly in contact.
- **Open pack ice** : Concentration 4/10 to 6/10, many leads and polynyas, floes generally not in contact.
- **Very open pack ice** : Concentration 1/10 to 3/10.

Ice free : No ice present. If ice of any kind is present, this term shall not be used.

Ice island : A large piece of floating ice protruding about 5 m above sea level, which has broken away from an Arctic ice shelf. They have a thickness of 30-50 m and an area of from a few thousand square metres to 500 km^2 or more. They are usually characterized by a regularly undulating surface giving a ribbed appearance from the air.

Ice jam : An accumulation of broken river ice or sea ice not moving due to some physical restriction and resisting to pressure.

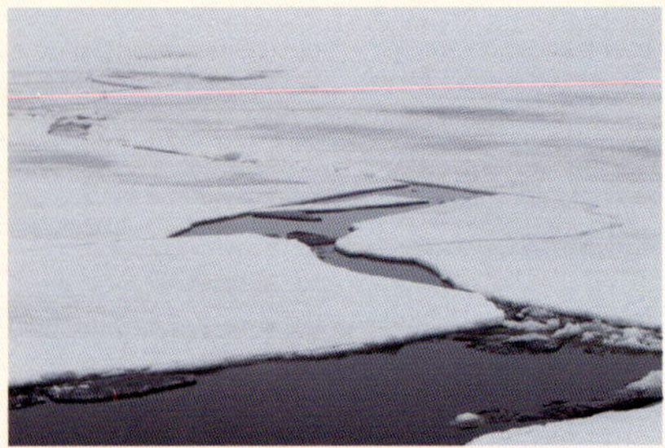

Ice keel : A downward-projecting ridge on the underside of the ice canopy; the submerged counterpart of a ridge. Ice keels may extend as much as 50 m below the surface.

Ice rind : A brittle, shiny crust of ice formed on a quiet surface by direct freezing or from grease ice, usually in water of low salinity. It has a thickness of about 5 cm. Easily broken by wind or swell, commonly breaking into rectangular pieces.

Ice shelf : A floating ice sheet of considerable thickness showing 2 m or more above sea level, attached to the coast. They usually have great horizontal extent and a level or gently undulating surface. Ice shelf growth occurs by annual snow accumulation and also by the seaward extension of land glaciers. Limited areas may be aground. The seaward edge is termed an ice front.

Iceberg : A massive piece of ice of greatly varying shape, protruding 5 m or more above sea level, which has broken away from a glacier and which may be afloat or aground. They may be described as tabular, domed, pinnacled, wedged, drydocked or blocky. Sizes of icebergs are classed as small, medium, large and very large.

Large iceberg : A piece of glacier ice extending 46 to 75 m above sea level and with a length of 121 to 200 m.

Lead : Any fracture or passageway through ice which is navigable by surface vessels.

Level ice : Ice unaffected by deformation.

Multi-year ice : Old ice which has survived at least two summer's melt. Hummocks are smoother than on second-year ice and the ice is almost salt-free. Where bare, this ice is usually blue in colour. The melt pattern consists of large interconnecting, irregular puddles and a well-developed drainage system.

New ice : A general term for recently formed ice which includes frazil ice, grease ice, slush and shuga. These types of ice are composed of ice crystals which are only weakly frozen together (if at all) and have a definite form only while they are afloat.

Nilas : A thin elastic crust of ice, easily bending on waves and swell and under pressure growing in a pattern of interlocking "fingers" (finger rafting). Nilas has a matte surface and is up to 10 cm in thickness and may be subdivided into dark nilas and light nilas.

Nip : Ice is said to nip when it forcibly presses against a ship. A vessel so caught, though undamaged, is said to have been nipped.

Open water : A large area of freely navigable water in which ice is present in concentrations less than 1/10. No ice of land origin is present.

Pancake Ice : Predominantly circular pieces of ice 30 cm to 3 m in diameter, up to 10 cm in thickness, with raised rims due to the pieces striking against one another. It may form on a slight swell from grease ice, shuga or slush or as a result of the breaking of ice rind, nilas or, under severe conditions of swell or waves, of grey ice. It also sometimes forms at some depth at an interface between water bodies of different physical characteristics where it floats to the

surface. It may rapidly form over wide areas of water.

Pinnacled iceberg : An iceberg with a central spire or pyramid, with one or more spires.

Polynya : Any non-linear shaped opening enclosed by ice. May contain brash ice and/or be covered with new ice, nilas or young ice; submariners refer to these as skylights.

Puddle : An accumulation of water on ice, mainly due to melting snow, but in the more advanced stages also to the melting of ice.

Rafted Ice : Type of deformed ice formed by one piece of ice overriding another.

Ridge : A line or wall of broken ice forced up by pressure. It may be fresh or weathered. The submerged volume of broken ice under a ridge, forced downwards by pressure, is termed an ice keel.

Second-year ice : Old ice which has survived only one summer's melt. Thicker than first-year ice, it stands higher out of the water. In contrast to multi-year ice, summer melting produces a regular pattern of numerous small puddles. Bare patches and puddles are usually greenish-blue.

Shuga : An accumulation of spongy white ice lumps having a diameter of a few centimeters across; they are formed from grease ice or slush and sometimes from anchor ice rising to the surface.

Slush : Snow which is saturated and mixed with water on land or ice surfaces or as a viscous floating mass in water after a heavy snowfall.

Small Iceberg : A piece of glacier ice extending 5 to 15 m above sea level and with a length of 15 to 60 m.

Snow drift : An accumulation of wind-blown snow deposited in the lee of obstructions or heaped by wind eddies. A crescent-shaped snowdrift, with ends pointing down-wind, is called a snow barchan.

Tabular iceberg : A flat-topped iceberg. Most show horizontal banding.

Weathering : Processes of ablation and accumulation which gradually eliminate irregularities in an ice surface.

Young ice : Ice in the transition stage between nilas and first-year ice, 10-30 cm in thickness. May be subdivided into grey ice and grey-white ice.

References

- Sanderson, T.J.O. 1988, Ice Mechanics - Risks to Offshore Structures, Granham & Trotman, London.
- Weeks, W.F. and Ackley, S.F. 1982, The Growth, Structure and Properties of Sea Ice, CRREL Monograph 82-1.
- KOPRI Homepage http://www.polar.re.kr

Index

Kyungsik Choi

During 2019 Antarctic Expedition

1980 Graduated from the Department of Naval Architecture, Seoul National University

1983 Acquired Master of Engineering, Department of Naval Architecture, Seoul National University (Major in Ship Structure)

1989 Acquired Ph.D. M.I.T. Dept. of Ocean Engineering

1989-Present Professor of Ocean Engineering at Korea Maritime and Ocean University

1992-1993 Participated in the 7th Summer Antarctic Research Group of KORDI

2010-2019 Four times participated in the Arctic and Antarctic Expedition onboard the Icebreaking Research Vessel ARAON

Major Research Areas : Polar engineering and Ice mechanics
Arctic sea route and icebreaker design
Ice load measurement and ice-structure interaction analysis
Polar pipeline and permafrost construction and design

Arctic and Antarctic Sea Ice

Published by Dasom Publishing Co.
10-1, Daecheong-ro 135beon-gil, Jung-gu, Busan, Republic of Korea
Tel : +82-51-462-7207, Fax : +82-51-465-0646, E-mail : hyosung2@chol.com

ISBN 978-89-5562-703-9 93450

Printed in Korea